PERGAMON INTERNATIONAL LIBRARY
of Science, Technology, Engineering and Social Studies
The 1000-volume original paperback library in aid of education, industrial training and the enjoyment of leisure
Publisher: Robert Maxwell, M. C.

MEASUREMENTS AND TIME REVERSAL IN OBJECTIVE QUANTUM THEORY

INTERNATIONAL SERIES IN NATURAL PHILOSOPHY

GENERAL EDITOR: D. TER HAAR

VOLUME 75

SOME OTHER BOOKS IN THIS SERIES
(*See also full list of related titles at the end of this volume*)

BUSCH, G. & SCHADE, H.
Lectures on Solid State Physics

PATHRIA, R. K.
The Theory of Relativity

AUVRAY, J. & FOURRIER, M.
Problems in Electronics

KOVACS, I. & ZSOLDOS, L.
Dislocations and Plastic Deformations

ROUSSEAU, M. & MATHIEU, J. P.
Problems in Optics

HAUG, A.
Theoretical Solid State Physics

TALPE, J.
Theory of Experiments in Paramagnetic Resonance

TAKAHASHI, Y.
An Introduction to Field Quantization

FAIN, V. M. & KHANIN, Y. I.
Quantum Electronics

KURTH, R.
Introduction to Stellar Statistics

DAVYDOV, A. S.
Quantum Mechanics

BILENKY, S. M.
Introduction to Feynman Diagrams

MEASUREMENTS AND TIME REVERSAL IN OBJECTIVE QUANTUM THEORY

by

F. J. BELINFANTE

PERGAMON PRESS

OXFORD · NEW YORK · TORONTO · SYDNEY · PARIS · BRAUNSCHWEIG

U. K.	Pergamon Press Ltd., Headington Hill Hall, Oxford OX3 0BW, England
U. S. A.	Pergamon Press Inc., Maxwell House, Fairview Park, Elmsford, New York 10523, U.S.A.
CANADA	Pergamon of Canada, Ltd., 207 Queen's Quay West, Toronto 1, Canada
AUSTRALIA	Pergamon Press (Aust.) Pty. Ltd., 19a Boundary Street, Rushcutters Bay, N.S.W. 2011, Australia
FRANCE	Pergamon Press SARL, 24 rue des Ecoles, 75240 Paris, Cedex 05, France
WEST GERMANY	Pergamon Press GmbH, 3300 Braunschweig, Postfach 2923, Burgplatz 1, West Germany

First edition 1975

Libary of Congress Cataloging in Publication Data

Belinfante, Frederik Jozef.

Measurement and time reversal in objective quantum theory.

(International series in natural philosophy; v. 75)

Includes indexes.

1. Quantum theory. 2. Wave function. I. Title. QC174.12.B44 530.1'2 74-32177

ISBN 0-08-018152-X

Printed in Hungary

CONTENTS

PREFACE

TEXTBOOKS on wave mechanics usually give an excellent account of how wave functions are used in ordinary nonrelativistic wave mechanics, but there are a number of questions about these wave functions (or *state vectors*, as they are often called) which usually are hardly discussed in them. Some of these questions are: "Do wave functions objectively exist in nature, like electric fields, or are they constructions of the mind, depending upon the whims of their user?" "What determines the state vector?" "Can the state vector be measured?" "Has each physical system its own state vector?" "If ψ predicts a probability distribution $|\psi(\mathbf{x})|^2$ and if I measure $\mathbf{x}$, so that the smeared-out probability distribution that was given by $|\psi(\mathbf{x})|^2$ suddenly changes by the measurement into a near certainty as might be described by a density that is a Dirac δ-function, how can $|\psi(\mathbf{x})|^2$ that abruptly be changed into a δ-function by the measurement?" "Can wave mechanics only predict future probabilities, or can it also "postdict" probabilities for what happened in the past?" and so on. These questions are important for a good understanding of what quantum theory has to offer us. Though specialists have been discussing these questions since long ago, the omission of their discussion from textbooks has created a generation of physicists often disagreeing among each other what answers these questions deserve. This has often led to apparent paradoxes, of which some of the best known ones are the Einstein–Podolsky–Rosen paradox, and the paradoxes about Schrödinger's cat and about Wigner's friend.

The Einstein–Podolsky–Rosen paradox is about an ensemble $E^{\mathbf{AB}}$ of pairs of systems **A** and **B** that interacted in the past. (An ensemble is an "infinitely large" imagined collection of objects that have some given properties in common. The systems here considered may be elementary particles, atomic systems, or other physical objects.) In particular, the ensemble considered by Einstein, Podolsky, and Rosen is such that it may be described by a state vector of the form $\psi^{\mathbf{AB}} = \sum_n c_n \psi_n^{\mathbf{A}} \psi_n^{\mathbf{B}}$. Let the $\psi_n^{\mathbf{A}}$ and the $\psi_n^{\mathbf{B}}$ be eigenfunctions of the operators $\mathcal{A}$ and $\mathcal{B}$ for observables A and B that can be measured upon **A** and **B**. Let $\phi_m^{\mathbf{A}}$ and $\phi_m^{\mathbf{B}}$ similarly be eigenfunctions of the operators $\mathcal{C}$ and $\mathcal{D}$ of observables C and D for the **A** and **B**. We shall assume that the operators $\mathcal{A}$ and $\mathcal{C}$ do not commute, and that the normalized eigenfunctions are related among each other by $\psi_n^{\mathbf{A}} = \sum_m a_{nm} \phi_m^{\mathbf{A}}$ and by $\sum_n c_n a_{nm} \psi_n^{\mathbf{B}} = k_m \phi_m^{\mathbf{B}}$. It then is easily seen that we may also write $\psi^{\mathbf{AB}} = \sum_m k_m \phi_m^{\mathbf{A}} \phi_m^{\mathbf{B}}$.

An example of this is an ensemble of pairs of spin $\frac{1}{2}$ particles in the singlet state. A and B then may be the z-components of the spins of **A** and of **B**, while C and D may be their x-components.

We now assume that each pair splits apart, and, after **A** and **B** are widely separated, we measure upon **A** either A or C, and on **B** either B or D. Quantum theory then predicts that for each original pair upon which we choose to measure A and B, we shall find $B = B_n$ whenever we find $A = A_n$, while, for each pair on which we measure C and D, we find the occurrence of $D = D_m$ similarly correlated to $C = C_m$. This may also be expressed by considering the subensembles E_n^A and E_m^C of $E^{\mathbf{AB}}$ which comprise those split pairs on which a measurement upon **A** yielded $A = A_n$ or yielded $C = C_m$. Quantum theory then claims that these subensembles are describable by state vectors $\psi_n^{\mathbf{A}}\psi_n^{\mathbf{B}}$ and $\phi_m^{\mathbf{A}}\phi_m^{\mathbf{B}}$. Therefore, for a measurement of B upon **B**, there will be in E_n^A a 100% probability of finding $B = B_n$, but in E_m^C there will be a probability distribution over various possible eigenvalues of B. Similarly, for a measurement of D upon **B**, a certain outcome ($D = D_m$) will be found in the ensemble E_m^C, and not in E_n^A.

Einstein and his collaborators felt this prediction of quantum theory to be paradoxical, as no signal from **A** telling which measurement (A or C) had been made could reach the distant **B** in time before the measurement there, in order to tell it to which ensemble it belonged. If upon all **B** the observable B is measured, this should give $B = B_n$ for all systems **B** in the ensemble E_n^A, but how would the system (or particle) **B** know that A and not C had been measured upon **A**, and that the result had been A_n? In classical physics, this is no problem. For instance, in the example mentioned above, if the spin were a classical property of the particles, we might say for pairs in a singlet state (total spin zero) that necessarily particle **B** would carry a spin opposite to particle **A**, explaining the correlation (with $B_n = = -A_n$ and $D_n = -C_n$). Since, however, S_z or S_x would be measured upon **A** *after* its separation from **B**, **B** would have to remember from the time it broke from **A** not only what value of S_z each particle took at the split, but also which value of S_x. In quantum theory, however, we cannot simultaneously have definite values for both S_z and S_x, so that it is impossible for particles **B** to carry from the time of separation from **A** the information needed for "explaining" the statistical correlations between **A** and **B** which quantum theory predicts.

Therefore, it is impossible to describe the quantum-mechanical predictions by the methods of classical physics; *i.e.* we should not try to understand the information which quantum mechanics provides us about the particles **B** in the ensembles E_n^A and E_m^C, by regarding the individual particles **B** as *carriers* of this information. There is a type of *hidden-variables* theories that assumes that the particles are the carriers of this information. In my book *A Survey of Hidden-Variables Theories* I call them "hidden-variables theories of the second kind." Such theories do predict correlations between the results of measurements upon systems **A** and **B** that interacted in the past. However, some of these predictions will have to disagree with the predictions made by quantum theory, and it is the task of experiments such as the ones performed by Freedman and Clauser, to decide which theory is correct; quantum theory, or some hidden-variables theory. I refer to the book mentioned for details. In the present book, we will only discuss the claims of *quantum theory*.

The Einstein–Podolsky–Rosen paradox is not a contradiction

between one quantum-mechanical claim and another one. It merely is a disagreement between quantum theory and attempts to assign, to an individual system like **B**, some properties which quantum theory denies it could have simultaneously.

In the example discussed by Einstein, Podolsky, and Rosen (EPR), and by Bohr in his reply, the A, B, C, and D were not spin components. Here, we used spins merely as another illustration of the problem. EPR's conclusion was that "quantum mechanics is incomplete." By this they meant that it cannot simultaneously describe properties $B = B_n$ and $D = D_m$ which EPR thought the particles **B** should possess already at the time they broke away from particles **A**, as they would later be found to have such properties, depending upon our choice of measurement of A or of C upon **A**.

Bohr answered that it was better to call this *complementarity* rather than incompleteness. We have, of course, the choice between measuring A or measuring C upon **A**; but, once we have chosen to measure A, we have lost our ability to predict D on **B**. (That is, D can have more than one value in the ensemble $E_n^{A\mathbf{B}}$ of systems **B** in the pairs **AB** that make up ensemble E_n^A.)

For each particle **B** in $E_n^{A\mathbf{B}}$, it is sometimes said that "this has $B =$ $= B_n$." It would be somewhat better to say that "this **B** is in a state $\psi_n^{\mathbf{B}}$," but even this is a contestable statement. We will not be led into contradictions if we make the watered-down claim that "for this **B**, *if* B is measured upon it, we shall find $B = B_n$." On the other hand, the statement "**B** is in state $\psi_n^{\mathbf{B}}$" ought to mean that "measurement of B upon this **B** will *with certainty* give $B = B_n$," and this claim would remain doubtful, because *it cannot be verified experimentally*. If we measure B and find $B = B_n$, this does not give any guarantee that this result was not a happenstance, just as any result $D = D_m$ would be a happenstance if we had chosen to measure D upon this **B**. In order to verify the *certainty* of $B = B_n$, we have to measure B on a representative sample of the ensemble $E_n^{A\mathbf{B}}$.

Therefore, we want to regard the state $\psi_n^{\mathbf{B}}$ as a property of the ensemble $E_n^{A\mathbf{B}}$, rather than of any one of the individual systems **B** that are part of $E_n^{A\mathbf{B}}$. When we assign $\psi_n^{\mathbf{B}}$ to $E_n^{A\mathbf{B}}$ and $\phi_m^{\mathbf{B}}$ to $E_m^{C\mathbf{B}}$, we are less likely to make EPR's error of trying to assign $B = B_n$ as a property to some individual particle **B**.

The paradox of Schrödinger's cat is about a cat in a room containing a capsule, filled with poison gas, which is automatically opened if during some predetermined time a certain quantum-mechanical process takes place for which ψ predicts, say a 36% probability. That is, we consider a case in which, at the end of this time interval, we have $\psi = 0.6\,\psi_D + 0.8\,\psi_L$, where ψ_D is the normalized wave function for the state in which the process took place and the cat was poisoned, while ψ_L is the state in which the cat remained alive. The question is: Is the cat dead, or is it alive, if we do not know whether or not that quantum process took place? Is the cat 36% alive and 64% dead? The latter assumption would seem to be ridiculous. Yet, some people have thought that it should be taken serious. The question, therefore, is apparently not entirely trivial. (It becomes trivial, though, when we regard ψ as giving statistical properties to an ensemble, and not as a property of one particular cat or its surrounding.)

The paradox of Wigner's friend is related to the one of Schrödinger's cat. It arises when with Wigner one considers the possibility that state vectors might be ascribed to individual systems, and that a state vector like the one for Schrödinger's cat should be interpreted as meaning that the question whether the cat is dead or alive is as yet undetermined. Since it is well known that upon observation the cat is found to be either 100% dead or 0% dead, Wigner then suggested that the act of *the observer becoming aware of the answer* to this yes–no question might somehow change the state $\psi = 0.6\,\psi_D + 0.8\,\psi_L$ into either ψ_D or ψ_L for the cat observed.

This kind of a change of ψ after a measurement, called the *reduction* of ψ, was proposed already in the early days of wave mechanics. The formula telling the result of this change is known as *von Neumann's projection postulate*. Bohm and Bub in 1966 have invented a hidden-variables theory (what I would call a theory "of the first kind") which attempts to describe this "collapse of the wave function" by non-linear terms added to the Schrödinger equation. Here is not the place to discuss this theory. It is up to experiments like the one performed by Papaliolios in 1967, to decide whether such a theory would give a better explanation of the observed facts than quantum theory does. So far, this does not seem to be the case.

In Chapter 1 we shall regard this sudden change of ψ as descriptive of a change in our choice of the ensemble which we want to describe by ψ. If, however, with Wigner we tentatively would assume that ψ would be a property of the object on which the measurement is made, this option of understanding the change of ψ is lost, and we must in that case look for another explanation.

For arriving at such an explanation, Wigner (W) discusses the paradox about his friend (F). W has asked F to perform for him a measurement upon some object (O), and, after F at time t_F has completed the measurement, W asks at time t_W what the result was. As W becomes aware of the result at time t_W, the change of ψ from a superposition (indefinite outcome) to a single term (describing a definite result) according to W takes place at time t_W; so that W must assume that, during the interval between t_F and t_W, the system (O+F) still was in a state of indefiniteness. On the other hand, when W later asks F how F felt about this matter during this time interval, F answers that already from t_F on he definitely knew that the outcome of the measurement upon O was a definite one.

This paradox is then resolved by Wigner by giving up the idea that the cause of the collapse of ψ must be the awareness by the observer himself. If already before t_W the object was in a state of definite result, apparently even for W, as observer of the (O+F) system, ψ should change to a state of definite result of the measurement upon O at the time when F becomes aware of this result. Thus rejecting the solipsism characterizing his original suggestion, Wigner made a new suggestion, that some kind of interaction between living minds and inanimate nature would change the state of O from an indefinite one to a definite one as soon as *any* mind would become conscious of the outcome of a measurement upon O.

Though Wigner has invoked the law of action and reaction for justifying this suggestion (remarking that, as our mind is influenced by our surrounding, it is only natural to assume that our surrounding will be influenced by our mind), most physicists have rejected Wigner's suggestion as too far-fetched, and as unnecessary, in view of the simpler interpretation of the reduction of ψ provided by "objective" quantum theory, which will be discussed in Chapter 1 of this book.

The time arrow of quantum theory is the subject of Chapter 2. As quantum theory in its conventional form predicts future probabilities from preceding information, it seems to grant a preference to one direction in time. Yet, couldn't we use the Schrödinger equation for calculating an earlier state vector from a later one? We may therefore ask whether quantum theory allows us to calculate probabilities for happenings in the past. Our Chapter 2, dealing with this question, is an extension of a paper written in 1964 by Aharonov, Bergmann, and Lebowitz.

In Chapter 3 we look back at our conclusions, and add a number of comments.

The purpose of this book was primarily to give a reply to scientific papers written, by a large number of authors, about fundamental questions like the ones mentioned above and some others. (For instance, we also discuss Everett's theory of the "branching universe.") This reply, however, may also serve as a supplement to textbooks which fail to present an appropriate discussion of these matters. We encourage students of wave mechanics to read our comments. Much of the above, as an introduction to the chapters to follow, was written with these students in mind, as most of the more advanced readers assumedly were already well acquainted with the paradoxes which I discussed.

The following chapters mainly use methods of nonrelativistic wave mechanics well discussed in introductory courses on wave mechanics. An exception, unfortunately, must be made about the theory of the **density matrix** and of **mixed states**, which I had to use several times. Therefore, those not familiar with these methods are urged to read first **Appendix G**, in which these matters are explained to the uninitiated.

One more comment should be made for a good understanding of the following chapters. When we want to measure a quantity A, our result would not be very meaningful if A would rapidly keep changing during the course of the measurement. Therefore, quantities measured are usually assumed to be (nearly) *conserved* quantities. When we say that the state ψ prepared by an ideal measurement yielding $A = A_n$ is the eigenfunction ϕ_n of the corresponding operator $\mathcal{A}$, we should perhaps have written (in a theory that uses the Schrödinger equation) that the state would be $\exp(\mathcal{H}t/i\hbar)\,\phi_n$. In order to avoid this complication, we would like to transform away most of this time dependence. Yet, we would like to keep in ψ enough time dependence for showing how the states of object and of apparatus of measurement change upon their interaction. For this purpose, we may somehow split up the hamiltonian $\mathcal{H}$ into a part $\mathcal{H}_I$ describing this interaction, and the remainder $\mathcal{H}_0$ which is the hamiltonian in absence of this interaction. We may then use $\exp(-\mathcal{H}_0 t/i\hbar)\,\psi$ as our new wave function. This eliminates the uninteresting part of the time dependence of ψ. This new ψ is called the *state vector in the interaction picture*. If we call it ψ_I, and write ψ_S for Schrödinger's time-dependent ψ, we have

$$
\begin{aligned}
\psi_\mathrm{I} &= e^{-\mathcal{H}_0 t/i\hbar}\,\psi_\mathrm{S},\\
i\hbar\,\partial\psi_\mathrm{I}/\partial t &= -\mathcal{H}_0\psi_\mathrm{I} + e^{-\mathcal{H}_0 t/i\hbar}\,i\hbar\,\partial\psi_\mathrm{S}/\partial t\\
&= e^{-\mathcal{H}_0 t/i\hbar}\,[-\mathcal{H}_0 + \mathcal{H}]\,\psi_\mathrm{S}\\
&= e^{-\mathcal{H}_0 t/i\hbar}\,\mathcal{H}_I\,e^{\mathcal{H}_0 t/i\hbar}\,\psi_\mathrm{I}\\
&= \mathcal{H}_I^\mathrm{I}\,\psi_\mathrm{I}.
\end{aligned}
$$

This again has the form of a Schrödinger equation, but $\mathcal{H}_0$ is missing from the hamiltonian. This Schrödinger equation of the interaction picture is sometimes called the *Tomonaga–Schwinger equation*, in particular when it is used in quantum field theory. We can now calculate the expected mean of an observable A from its operator by

$$\langle A\rangle = \int \psi_\mathrm{S}^\dagger\,\mathcal{A}\,\psi_\mathrm{S} = \int \psi_\mathrm{I}^\dagger\,\mathcal{A}^\mathrm{I}\,\psi_\mathrm{I},$$

if in the interaction picture we use the transformed operators

$$\mathcal{A}^{\mathrm{I}} = e^{-\mathcal{H}_0 t/i\hbar}\,\mathcal{A}\,e^{\mathcal{H}_0 t/i\hbar}.$$

The interaction operator $\mathcal{H}_I^{\mathrm{I}} = \mathcal{H}^{\mathrm{I}} - \mathcal{H}_0^{\mathrm{I}}$ which in the Tomonaga–Schwinger equation takes the place of the hamiltonian operator is, therefore, obtained from the transformed hamiltonian by subtracting $\mathcal{H}_0^{\mathrm{I}} = \mathcal{H}_0$ from it.

In the following chapters, we do not write indices I. The reader may insert them, if he wants to consider our presentation to be made in the interaction picture rather than in the Schrödinger picture. Where interactions are absent ($\mathcal{H}_I = 0$, $\mathcal{H} = \mathcal{H}_0$), we may equally well use the *Heisenberg picture:*

$$\mathcal{A}^{\mathrm{H}} = e^{-\mathcal{H}t/i\hbar}\,\mathcal{A}\,e^{\mathcal{H}t/i\hbar}.$$

Examples of this are found in Appendices **A** and **B**.

ACKNOWLEDGEMENTS

I GRATEFULLY acknowledge conversations and correspondences with several colleague physicists. My attention to disagreements among workers in this field was first drawn by papers written by L. E. Ballentine and by P. A. Moldauer. A paper by B. S. DeWitt drew my attention to the work of H. Everett III. An unknown referee of a first (unpublished) note which I wrote on the reduction problem drew my attention to the books by B. d'Espagnat. My Purdue colleague H. M. James was my sounding board for some early oral discussions. H. Margenau asked my opinion about some of his own publications, including his work with J. L. Park. The work of Y. Aharonov, P. G. Bergmann, and J. L. Lebowitz taught me that quantum theory becomes incomplete when its asymmetry in time is formally removed. L. Rosenfeld honored me with an extended correspondence on some points of this work. I am very grateful to E. P. Wigner for several oral discussions: at his home, at conferences, and over the phone. I hope these may have somewhat reduced my misunderstandings about some of his ideas and those of d'Espagnat. Finally, I owe D. ter Haar for an elaborate list of suggestions for improvements in style and language.

I acknowledge financial support of this work from grant GP-29786 of the National Science Foundation.

CHAPTER 1

MEASUREMENTS IN OBJECTIVE QUANTUM THEORY

1.1. Introduction

The basic elements of quantum theory consist of two parts. On the one hand, it describes "pure states" by state vectors ψ (often called *wave functions*) of which the time dependence is given by a Schrödinger equation $i\hbar\partial\psi/\partial t = \mathcal{H}\psi$. On the other hand, there is a set of rules which tells us how these ψ physically are to be interpreted. Without the one or the other, the theory is incomplete. If the one or the other is fundamentally and drastically altered, we may still have a theory, but it would no longer be quantum theory. Whether the new theory would be better or would be worse than conventional quantum theory would have to be decided by experiment. If the new theory would not be experimentally distinguishable from quantum theory, the simpler of the two theories would be preferred by most people. (Occam's razor.)

The "rules" for interpreting ψ are well known. Each ψ predicts a probability distribution over possible outcomes of successful measurements of arbitrary observables A. The (theoretically) possible outcomes are the eigenvalues A_n of the operators corresponding to these observables. If ψ is expanded in terms of the corresponding eigenfunctions ϕ_n of such an operator, the absolute squares $|c_n|^2$ of the expansion coefficients are interpreted as probabilities for these outcomes of successful measurements of A.

Physical states more often are *mixed* states than *pure* states. A mixed state may be described by a classical probability distribution over pure states. If p_α is the probability for the pure state ψ_α which may be expanded in terms of a complete orthonormal set of functions $\{\phi_n\}$ by $\psi_\alpha = \sum_n c_{\alpha n}\phi_n$, then the expected mean value $\bar{A}$ of an observable A is given by $\bar{A} = \sum_\alpha p_\alpha \int \psi_\alpha^* \mathcal{A} \psi_\alpha =$ Trace $\{\mathcal{A}\mathcal{W}\}$, where the "density operator" or "density matrix" $\mathcal{W}$ is defined by[1] $\int \phi_m^* \mathcal{W} \phi_n = W_{mn} = \sum_\alpha c_{\alpha m} p_\alpha c_{\alpha n}^*$. In this case, W_{NN} is interpreted as the probability that successful measurement of A will yield the value A_N, if for $\{\phi_n\}$ we take the eigenfunctions of $\mathcal{A}$.[2]

In the early days of quantum theory, these rules were postulated or were in some way axiomatized. More recently, people have asked themselves whether the "rules" are compatible with the Schrödinger equation. If the rules are valid, the outcome of an ensemble of successful measurements could be described as a mixed state with a probability distribution over state vectors each describing the subensemble in which one particular outcome is found. On the other hand, the initial state of the entire ensemble before the measurement may have been a pure state, and one can prove that the Schrödinger equation does not cause transitions of pure states into mixed states. This creates the so-called *measurement problem.*

Many authors have tackled this problem, and have come up with many mutually different verdicts. Some people have proposed to replace quantum theory by other theories. These theories have their own problems. For instance, various mutually different hidden-variables theories deal each in their own way with their measurement

[1] Conversely, the interpretation of a given density matrix W_{mn} as a probability distribution over state vectors is not unique. For instance, for electrons of spin $\frac{1}{2}\hbar\boldsymbol{\sigma}$, if $\mathbf{p}$ is a vector in the first octant such that $p_x+p_y+p_z = 1$, and if $\mathbf{N} = \mathbf{p}/|\mathbf{p}|$ and $\sigma_N = \mathbf{N}\cdot\boldsymbol{\sigma}$, the density matrix of a mixture of three pure states $\sigma_x = +1$, $\sigma_y = +1$, and $\sigma_z = +1$ with probabilities p_x, p_y, and p_z is identical with the density matrix of a mixture of the two orthogonal pure states $\sigma_N = \pm 1$ [density matrices $W_{N\pm} = \frac{1}{2}(1 \pm \sigma_N)$] with probabilities $p_{N\pm} = \frac{1}{2}(1 \pm |\mathbf{p}|)$. See also Sec. 2.3.

[2] If A_N is a degenerate eigenvalue of $\mathcal{A}$, W_{NN} should bereplaced by a sum of W_{nn} over all states in the set $\{\phi_n\}$ that belong to that same eigenvalue A .

problem, and in some of them the problem is still partially unsolved. We shall not discuss such theories here. I discussed them elsewhere.[3]

Some people have proposed reinterpretations of quantum theory, or, in their discussion of what they call quantum theory, they assign to ψ meanings which do not jibe with the conventional rules of quantum theory at all. These people actually are discussing *altered* theories, without explicitly saying so. This sometimes leads to controversies with those to whom quantum theory is characterized by the conventional rules, and is most confusing to those who do not specialize in this field and who do not realize that those so-called quantum-theoretical considerations are actually statements about some imagined theory that is basically different from conventional quantum theory.

In the following, we shall mainly deal with *conventional* quantum theory. Our ultimate purpose here is to explain (a) why this theory *allows* "reduction of the wave function" after a successful measurement, (b) to what extent reduction is merely *optional*, and (c) why it is usually wise to *make use* of this option.

It is *not* our purpose here to explain the details of *competitor* theories, nor is this a treatise on the various philosophies attempted by various people to apply to physics. For a discussion of quantum theory *versus* hidden-variables theories, see reference 3. For a discussion of various philosophical attitudes, see d'Espagnat's books.[4, 5] Quantum theory, and the laws of physics for which it stands, are in the present discussion taken as granted. We then confine ourselves to that philosophy of physics which *agrees* with these assumptions. As we go along, we shall try to make this choice plausible, without aiming at excessive rigor.

As conventional quantum theory maintains that the result of successful measurements is describable as a mixed state for the ensemble of

[3] F. J. Belinfante, *A Survey of Hidden-Variables Theories*. (Pergamon Press, Oxford, 1973.)

[4] B. d'Espagnat, *Conceptions de la physique contemporaine*. (Hermann, Paris, 1965.)

[5] B. d'Espagnat, *Conceptual Foundations of Quantum Mechanics*. (W. A. Benjamin, Menlo Park, California, 1971.) This book, with a six-page but incomplete list of printing errors, is *not* a translation of the first one.[4]

objects of the measurement, and as it can be shown that a pure state of these objects in interaction with apparatus cannot become a mixed state by the Schrödinger equation, how can we justify that these two statements do not contradict each other? Therefore: How do we draw conclusions, about the state of the ensemble of the *objects alone after* successful measurement, from the state of the ensemble of systems of *interacting object and apparatus during* the measurement? Are object and apparatus separable? Are our conclusions made possible by the *successfulness* of the measurement? Are there restrictions to the applicability of the conventional rules of quantum theory?[6] These and related questions will be discussed in the later part of this Chapter 1.

Before we go into this, we shall briefly review some of the fundamentals of conventional quantum theory, in order to avoid misunderstandings about the nature of the theory we are discussing. As, however, discussion of some salient points is postponed until the later sections of this chapter, we want to caution the reader against drawing his conclusions too soon.

As an example, consider the following problem of statistical mechanics. (This example has also some bearing upon things we will discuss in Chapter 2.)

Suppose we have a vessel with its left-hand half filled with hydrogen gas and its right-hand half filled with fluorine gas. (This is state A.) Soon after, we will have HF everywhere. (This is state B.) The transition $A \rightarrow B$ is *irreversible*. Nevertheless, the equations of motion involved (classical ones or a Schrödinger equation) are time-reversible. From the latter fact, one might easily conclude that *irreversible processes do not exist*. This conclusion apparently is drawn too soon, as is well demonstrated by the occurrence of the transition $A \rightarrow B$, and the absence of transitions $B \rightarrow A$ at the same temperature.

The solution of this paradox is well known. State A can easily be prepared directly, by first filling the two halves of the vessel while a separating wall is in place, and by then quickly pulling away the separator. It is, however, practically impossible to prepare state B with the internal positions and motions of all particles prepared in such a way that a process $B \rightarrow A$ will take place that is the time-reversed of any of the possible processes $A \rightarrow B$. Therefore, even *if* such a state B would

[6] Wigner[7] has stressed already in 1952 that the conventional mathematical formulation of the theory of measurements is an approximation valid only if the apparatus can take up certain types of recoil without being noticeably perturbed by it. The use of macroscopic instruments for measuring microscopic quantum effects is a method for satisfying this condition.

[7] E. P. Wigner, *Zeitschrift für Physik*, **133**, 101 (1952).

exist theoretically, we may as well forget about it, as it will never be realized or used as a method for realizing A.

If physics is to be a discipline that we can apply to our dealings with nature, therefore, we have to be careful not to oppose the *mathematical* statement that time-reversible equations of motion do not allow time-irreversible elementary processes, to the *physical* statement that thermodynamically irreversible processes are common-place.

Similarly, it is somewhat hasty to conclude from the impossibility of a pure state of an ensemble of isolated systems to go over automatically into a mixed state of the same, that it would be impossible to make successful measurements upon those systems.

1.2. Topics to be discussed

In the following sections, we shall first explain why quantum theory may be regarded as a theory about *ensembles* rather than about *elementary systems*. In this context, we shall caution the reader against a certain use made by philosophers of the word *realism* in connection with physical theory. (Sec. 1.4.)

Next, we ask what are the elements (*members*) of the ensembles discussed by quantum theory.[8] To what extent is quantum theory *deterministic*, and to what extent is it *not*? In what sense can a state vector sometimes be assigned to an *individual* system, and to what extent is *even that* state vector really a state vector for an *ensemble*? Why is quantum theory an *objective*[9] discipline? How does the state vector correspond to a *method of preparing* a system? Why does it *not* necessarily describe our *knowledge* of nature, and how do we keep a discussion of *consciousness* out of quantum theory? How does *Schrödinger's cat* characterize a misunderstanding of quantum theory, from our conventional (objective) point of view?

Why is reduction of ψ a *privilege* and *not a necessity*? Why is reduction *not* a change of mind about the ψ for an *unchanged object*, but merely a change of our choice of the object of our discussion? Why is it meaningless to talk about the *time* at which reduction "takes place"? Discussion of these questions brings us to Sec. 1.16.

[8] This brings up the question of *separability* of systems, which is postulated in the first part of this chapter, and which is discussed in its later part.

[9] The objective nature of quantum theory was brought to my attention by Niels Bohr in oral discussion.

After these preliminary considerations, we look at the usual *schematical mathematical description of measurements*. This leads to the *measurement problem*, which shows the need for a *justification* of at least *qualified* validity of the probability interpretation of ψ. We find a direction in which this justification can be obtained by discussing a problem considered already by Einstein and Bohr. An example discussed in the past by Wigner leads to the same conclusions. We generalize these ideas by using a mathematical schematization conveniently introduced by Everett,[10] utilizing a so-called *memory state vector*. This provides a more definite description of what is a *successful* measurement.

Everett himself used these memory state vectors in his *theory of relative states*, which was supposed to completely *eliminate the need for reduction of state vectors* even in calculations of probabilities for follow-up measurements. His interpretation of quantum theory is said to describe a *branching of the universe*,[11] and therefore is rejected by most people.[12] We show how Everett's memory state vectors may just as well be used for *justifying* the reduction procedure. Thus, the question of reduction would seem to become a question of taste. We point out, however, that description of nature by ψ would be a hopeless task if Everett would not smuggle in a reduction of state in the description of any *initial state* for a quantum-mechanical problem. Then, it becomes illogical not to use the same simplifying method for the

[10] H. Everett, III, *Revs. Modern Phys.* **29,** 454 (1957). In 1973, a more elaborate version of this work (said to have been written by Everett already in 1955) has been published in book form: *The Many-Worlds Interpretation of Quantum Mechanics*, edited by Bryce S. DeWitt and Neill Graham (Princeton University Press, Princeton N. J., 1973). This presentation of Everett's work admits the validity of Bohr's interpretation of quantum theory without discussing it. However, it calls the Bohr interpretation "overcautious" (page 111), and prefers Everett's theory of a universal wave function ψ. This ψ is supposed to describe "the entire universe," instead of an ensemble of universes as it does in the version of it which we shall present in Secs. 1.21–1.27 below. Since we there merely discuss this modified interpretation of Everett's theory, our objections against it discussed in Sec. 1.26 may be regarded as objections against this reinterpretation, but in that case our objections against Everett's original interpretation are contained in Secs. 1.3 and 1.12.

[11] See Sec. 1.25.

[12] See Secs. 1.26 and 1.27.

initial state of a follow-up measurement, and the reduction procedure finds its justification.

The advantage of Everett's method here lies in its *schematization of the reasons why the conventional interpretation of ψ is acceptable*, without delving into these reasons as others have done.

The justification of Everett's schematic memory state vector lies in the irreversibility of the mechanism which makes a measurement successful by creating an indelible record of the result of the measurement. In Chapter 2, on retrodiction, we will touch this question of irreversibility another time, there taking for granted the allowability of the reduction procedure.

Many of the points discussed in this Chapter 1 have been discussed by others in the past. A list of literature references is given in a paper by Ballentine.[13, 14] A survey of related literature was also made by DeWitt and Graham,[15] while a list of earlier publications was compiled by Margenau.[16] What is new in my presentation is mainly my utilization of Everett's memory state vectors for a purpose opposite to the purpose which Everett had in mind.

Also, I hope by the above and by the following considerations to clarify several points that have confused many, because in the literature these points more often have been taken for granted tacitly, than explained explicitly to disbelievers. For instance, the statement that ψ describes an ensemble, obvious without explanation to many, is at the basis of much of the following, and therefore does require a few paragraphs of explanation in view of the tenacity with which others have tried to adhere to the idea that state vectors ψ should describe individual elementary systems. As these persons usually have found that this interpretation of ψ leads to difficulties in understanding quantum theory, arguments in favor of the conventional interpretation of ψ as describing ensembles ought to be welcome to them as a way out of their difficulties.

That this easier interpretation "does not explain everything" is obvious. Quantum theory *does not claim* to explain everything. It merely claims to *describe* and *predict* the behavior of *ensembles* of elementary systems. If it could explain or predict what in the one universe we know exactly happens to one individual

[13] L. E. Ballentine, *Revs. Modern Phys.* **42,** 358 (1970).

[14] While we agree with much of Ballentine's paper, see Appendix B for a criticism of some of his chap. 3. In addition, what Ballentine calls the "statistical interpretation" has been the Copenhagen point of view since long, and what Ballentine calls the "orthodox" point of view should perhaps be called the "Von Neumann" point of view, but it *is not* "the" Copenhagen point of view (if it ever was).

[15] B. S. DeWitt and R. N. Graham, Resource Letter on the Interpretation of Quantum Mechanics, *Am. J. Phys.* **39,** 724 (1971).

[16] H. Margenau, *Annals of Physics (New York)*, **23,** 469 (1963).

elementary system we meet, the theory would be deterministic like a hidden-variables theory,[3] and would be entirely different from the conventional theory which we want to discuss here.

1.3. Probabilities and ensembles

Quantum theory provides a scheme for calculating probabilities from state vectors. The word "probability" has mainly two meanings. In the sentence "There is a great probability that John will find fault with this," "probability" is used as a qualitative property of a single system (John), and indicates something like an expected tendency. In the more scientific sentence "The probability that nonsmokers will be irritated when you smoke in their presence is great," the word "probability" is applied to a large group of people, and means "relative frequency" among these people, and one can assess an approximate numerical value to it. When in quantum theory we assign exact values to probabilities, we consider the idealized case of an "infinitely" large collection of cases among which a relative frequency is calculated. The theory, therefore, deals with *ensembles*. State vectors determine the probability distributions in ensembles: *State vectors are properties of ensembles*, and *quantum theory is a theory about properties of ensembles*.[17, 18] In Sec. 1.8 we shall deal with an apparent exception to this rule, but will find our present statement to keep its validity.

[17] Ballentine[13, 14] calls this the "statistical interpretation" of quantum theory, and warns against different interpretations. Those so-called different interpretations of the theory really are fundamentally different theories. Therefore, I will feel free to refer to such theories in Sec. 1.4 under a different name, and reserve the name "quantum theory" for a theory assigning numerical probabilities to happenings in ensembles.

[18] In his well-written paper *Quantum Mechanics of Individual Systems*, Hartle[19] concludes on page 708: "A quantum-mechanical state is ... not an objective property of the individual system." Rather, it describes the state of an ensemble of systems "prepared" in apparently identical ways. Applied to an individual member of the ensemble, therefore, the state vector makes predictions only when the probability in the ensemble is either zero or 100%. See Secs. 1.7 and 1.8.

[19] J. B. Hartle, *Am. J. Phys.* **36**, 704 (1968).

1.4. "Realistic" theories of elementary systems

By "elementary systems" I refer to systems at the atomic or subatomic level.[20] Quantum theory can apply only to *ensembles* of such systems, in as far as it is a theory of probabilities.[21] Many authors, however, have not been satisfied with this. They want a theory applicable to individual elementary systems.

As it is quantitatively meaningless to talk about probabilities for a single system,[21] theories of this kind must be deterministic, or at least cryptodeterministic. That is, if the system's behavior is not knowable explicitly, it should then at least be determined by some unknown parameters. If for the latter a stochastical distribution is proposed, we are back at considering a theory of ensembles, but, this time, at least in principle each single system in the ensemble would be individually characterizable by these "hidden" parameters.

When people want a theory of elementary systems, and yet they claim that they want to maintain quantum theory, they obviously are contradicting themselves.

Some of these people are driven toward theories of elementary systems by their desire that the theory should satisfy the "doctrine of realism" according to which observables defined on elementary systems shall "have" definite (known or unknown) values irrespective of whether anybody would want to measure them or of whether any interactions would take place that would depend upon this value. As explained in Sec. 2.13, and as easily grasped from the unsolved difficulties met by the abundant literature trying to assume such a "realistic" doctrine, this kind of realism does not agree with the fundamentals of quantum theory,[22] so that we must conclude that quantum theory is definitely what these people would call an "unrealis-

[20] This does *not* exclude macroscopic systems, provided we regard the latter as built up out of atomic systems. Also, light and fields are included, when described quantum-mechanically.

[21] See Sec. 1.12.

[22] Compare the last sentence of Sec. 1.5.

tic" theory, as *it does not deal with elementary systems themselves*,[23] but with ensembles of them.

The most logical alternative to quantum theory then would be some consistent[24] hidden-variables theory.[3] Theories of this kind thus far constructed, however, seem to disagree[25] with the results of recent experiments,[26, 27] though the question has been raised whether experimental results do not also disagree with quantum theory.[28] Until the latter disagreement has been confirmed, however, we shall here tentatively assume that quantum theory is corroborated by the observed facts.

In that case, the only thus far successful theory would remain quantum theory. Then, the doctrine of "realism" that favors "theories of elementary systems" would in fact be *unrealistic* in the sense that it would disagree with the only existing consistent interpretation of the empirical observations. It then might be better to call this doctrine "*pseudo*realism," and its avoidance in quantum theory would become a merit instead of a liability.

Even if Holt's results[28] would be correct and not caused by experimental errors, this would not affect the objective consequences of quantum theory as it stands, irrespective of its validity. While it is

[23] Except in cases where a certain probability happens to be 0% or 100%. See Sec. 1.8 for a discussion of this apparent exception.

[24] Various authors have proved the inconsistency of certain types of hidden-variables theories, which I call "hidden-variables theories of the zeroth kind." (See Part One of reference 3.) When I mention hidden-variables theories in the text above, I do not include these. In the theories to which I do refer, the outcome of any well-defined measurement is determined for any single elementary system **S** when we know for **S** a complete set of hidden variables in addition to the state vector ψ. The ensembles of quantum theory then correspond to probability distributions over hidden variables for systems **S** that have the same ψ in common.

[25] See reference 3, Parts II and III.

[26] C. Papaliolios, *Phys. Rev. Letters*, **18,** 622 (1967).

[27] S. J. Freedman and J. F. Clauser, *Phys. Rev. Letters*, **28,** 938 (1972).

[28] Experiments by R. A. Holt at Harvard University gave results which contradict the predictions of quantum theory. (See reference 3, page 290, footnote 25b.) As it is hard to understand why quantum theory would not hold for $j = 1$ states while it would hold for $j = 0$ states,[27] the experiments of Holt and those of Freedman and Clauser require confirmation by experiments performed elsewhere before they can be accepted as conclusive evidence.

hoped that further observations will either disprove Holt's results, or will help us find a new theory, better than quantum theory and explaining all experimental results, we shall here not further discuss this question, and, from here on, we shall merely discuss objective quantum theory as it is. Then we are forced to accept the statistical interpretation that the theory makes rules for *ensembles* of elementary systems, and in general[23] not for individual elementary systems.

1.5. Ensembles of what?

The rules of quantum theory usually are formulated for ensembles of *closed* systems. In nature, closed systems do not exist. However slight, there always is some exterior interaction, incident cosmic radiation, or outgoing radiation. Therefore, like any physical theory, quantum theory deals with *idealizations* in which certain things purposely are neglected for enabling us to deal effectively with the problems at hand.

As we will discuss in Sec. 1.25, a system's separation from the outside world is more of a problem in quantum theory than it is in classical theory. Anyhow, we must start by assuming that somehow this problem can satisfactorily be "solved" (or talked away), before we can hope to apply quantum theory at all to limited physical systems.

The problem mentioned arises immediately, as quantum theory discusses correlations between results of successive ideal measurements, or between a preparation of an initial state and the result of a following nonideal measurement. The object **O** of these measurements then interacts first with an apparatus **A** for measuring an observable *A* on **O**. Next, it interacts with an apparatus **B** for measuring the observable *B* on **O**. If **A** and **B** yield results that are registered in the memory **M** of some computer, the smallest closed system which can describe the happenings consists of (**O** $\oplus$ **A** $\oplus$ **B** $\oplus$ **M**). [From here on, we shall denote such a combination of systems simply by **OABM**.] Nevertheless, the conventional postulates of quantum theory predict results for an ensemble of such sequences of measurements in terms of (the ensemble of systems) **O** alone. This obviously can only be an approximation, of which it is postulated that it is a "good" approximation.[29]

The so-called "theory of measurements" schematically takes **A**, **B**, and possibly **M** into account. It then is shown how **A**, **B**, and **M** may be eliminated from the conclusions.[29] This leads to the conventional rules that do not mention **A**, **B**, or **M** at all.[30]

Again, those who are enslaved to the doctrine of (pseudo?) realism will here find it difficult to follow the methods of quantum theory, as quantum theory creates a conflict[31] between (pseudo) realism and the separability of systems that interacted in the past, like **O** and **A** in the above case.

1.6. Determinism for ensembles

Quantum theory maintains that there exists a causal law determining the evolution of the state vector ψ in time. In the present form of the theory, this law is the Schrödinger equation (or possibly a Tomonaga–Schwinger equation) $i\hbar\, \partial\psi/\partial t = \mathcal{H}\psi$. It supposedly is linear in ψ, thus ensuring the validity of the superposition principle often tacitly used throughout the following.

1.7. Indeterminism for elementary systems

Quantum theory claims that in general (with exceptions discussed in Secs. 1.8 and 1.10 below) no definite predictions can be made about a single elementary system. (The *apparent* predictability of macroscopic

[29] *All* physical theories are approximations in which something (knowingly or unknowingly) is being neglected. This sets limits to the validity of physical theories. Therefore, mathematics is in physics a useful tool *only when wisely handled*. When blindly applied to extreme cases outside the limits of validity of the theory, one should not be suprised if it leads to conflicts with the real facts or to absurdities.

[30] In Chapter 2 we shall use these conventional rules from the start.

[31] See, for instance, pages 25–28 and 42–47 of d'Espagnat's 1965 book,[4] or pages 111–124 of his 1971 book.[5] Readers of these two books are warned to read them *critically*, as d'Espagnat often discusses successively contradicting points of view and does not always clearly take sides, so that not all he discusses can be regarded to be his personal opinion. At times, this may confuse the reader.

systems is seen as a consequence of the determinism for ensembles and the smallness of the unpredictable deviations from the predicted averages at the atomistic level.) This indeterminism forces quantum theory to confine itself to *probability* predictions for *ensembles*.[23]

1.8. Eigenstates of conserved observables

An exception to unpredictability occurs when a predicted probability happens to be 100% (or if it is 0%). Thus, if a certain observable A for a system[32] **S** is conserved from time t_1 till time t_2, and if from an ideal measurement at t_1 we *know* its initial value (or, for keeping "knowledge" out of the description: if at t_1 some experimental setup *prepares* the systems **S** to have a certain initial value for the conserved observable), then measurement of this observable at t_2 ($> t_1$) will "with certainty" find "the same" value of that observable again.

(The quotation marks here remind us of unavoidable human errors, observational inaccuracies and the like, which are not described by the theory, and which in the following will be ignored by omitting further quotation marks for them.)

In the case mentioned, **S** may be considered a member of an ensemble E of similarly prepared systems for which the state is described by a state vector ψ which is an eigenfunction of the quantum-theoretical operator $\mathcal{A}$ for the observable. While ψ predicts the value of A for an immediately following measurement of it on the system **S** as well as on any other member of E, this ψ predicts for an immediately following measurement of an observable B not commuting with A (*i.e.* whose *operator* $\mathcal{B}$ does not commute with $\mathcal{A}$) merely a probability distribution in E.

In the case considered, the state vector ψ is by many authors ascribed as a property to the individual system **S**. From the fact that, about the next measurement of B, ψ predicts nothing for **S** alone, but predicts something interesting for E, we conclude that "ascribing ψ to **S**" really should be understood as ascribing ψ *to the ensemble E* which

[32] The system **S** here may, in the notation of Sec. 1.5, be **O** or **OA** or **OAM** or any other combination.

in this case is determined by the preparation of the state of **S**. The meaning which ψ has for **S** itself is only *part* of the meaning of ψ, and is anyhow *included* (through 100% and 0% probabilities for A values) in the meaning of ψ for E.

Therefore it is wise to maintain the notion of Sec. 1.3 that ψ describes the state of an *ensemble*, even when we encounter a system for which by ideal measurement or by preparation "the quantum state is entirely known" (but for a phase factor).

1.9. Objectivity

Quantum theory is objective[33] in as far as its *rules* are objective and prescribe the same theoretical conclusions irrespective of who makes them. That is, two persons considering *the same ensemble* will calculate the same probability distributions in it.

This does *not* mean that two different persons would always regard a given system **S** as a member of one and the same ensemble E. In fact, one single person may want to consider **S**—possibly even simultaneously—as being a member of more than one ensemble E. Thus, using state vectors ψ_1 and ψ_2 (or, more generally, density operators W_1 and W_2) for describing different ensembles E_1 and E_2 of both of which **S** simultaneously is a member, different probabilities w_1 and w_2 for a particular occurrence may be predicted simultaneously. There is nothing contradictory in this. If the happening would occur n_i times in ensemble E_i with N_i members, we have $w_i = n_i/N_i$. Even if n_1 and n_2 were equal (which may not necessarily be the case), $N_1 \neq N_2$ would make $w_1 \neq w_2$.

We therefore consider quantum theory to be an *objective*[33] theory of *ensembles*. Objectivity here merely means that w_i in ensemble E_i, as calculated from the density operator W_i for E_i, has a value independent of who calculates it. Note that an undetermined constant complex

[33] We use the word "objective" here in the "weak" sense in which it has often been used by Bohr.[9] Hartle[19] and our footnote 18 use the word "objective" in a different (stronger) meaning, as "experimentally verifiable and independent of the system's previous history."

phase factor in ψ_i affects neither W_i, nor predictions made. But for this phase factor, the ψ describing a given pure-state ensemble is objective.[33]

The "objective" point of view here taken is part of what Groenewold[34] has called the "skeptical" interpretation of quantum theory.

1.10. Reality of nature

Quantum-theorists do not deny that nature "really exists." What is meant by the latter is merely that under normal circumstances (in absence of anything like hallucinations) all persons observing an event appear to be able to come to an agreement about what was observed. This is said to "prove" that the observed fact has "reality." (Without this reality, the parenthesized second paragraph of Sec. 1.8 would lose its meaning.)

In particular, "when a system **S** is in an eigenstate ψ of certain observables" (which in quantum theory is an elliptical way of saying "when **S** has been prepared to be a member of an ensemble E described quantum-theoretically by the eigenstate ψ"), all physicists should agree about this, and it is permissible to say that the system **S** is "prepared to have" the corresponding (eigen) values of these observables.

Note, however, that it is incorrect here to omit the words "prepared to." The reason why this is so is explained in the last part of Sec. 2.11. Therefore, our admission of the reality of nature does *not* mislead us into accepting the doctrine of "realism" of the kind discussed in Sec. 1.4.

The "reality" which we do accept is not in conflict with the statement of Sec. 1.9 that a system **S** can simultaneously be a member of more than one ensemble. For instance, if measurement of A upon some **S** in E finds **S** to be a member of the subensemble E_N ($\subset E$) of systems for which $A = A_N$, *all observers* of this measurement would agree that **S** is a member of E_N as well as of E.

[34] H. J. Groenewold, *Physics Reports*, **11** C, 327 (1974).

1.11. Quantum theory and knowledge

Knowledge is important for making wise *use* of an existing theory. Let us see *how* we may use it.

Let z be an eigenstate of an observable Z, and let E_z be an ensemble of systems **O** prepared by selection of the outcome z of ideal measurements of Z upon the **O**. In this ensemble E_z, according to the assumptions of quantum theory, there is a probability $|\langle a|z\rangle|^2$ for finding the result a for a following measurement of A.[35] That is, a fraction $|\langle a|z\rangle|^2$ of the members of E_z form the subensemble E_{za} of systems **O** on which the measurement of A yields a.

In conventional quantum theory, this assumption is often stated as a *postulate* or as an *axiom*. Ballentine lists it as *rule* **F5** in chap. 1 of his paper.[13]

This axiom or rule implies that the successful completion of the measurement of A must have changed the state of the ensemble E_z of systems **O** into a mixed state.[36]

If, now, we happen to have *knowledge* of the result of one particular execution of the measurement A upon some particular system **O**, and we happen to know that our measurement was an ideal one, so that its repetition upon the same system **O** would certainly yield the same result, we may want to treat this system **O**, in the discussion of subsequent measurement of an observable B upon it, as a member of the subensemble E_{za} instead of merely as a member of E_z. Quantum theory postulates the predictive[37] state vector of E_{za} to be a, that is, the same[38] as for the ensemble E_a. (The previous state z of the ensemble

[35] As in Chapter 2, we assume here for simplicity nondegenerate observables. Otherwise, for a one should take the projection of z onto one of the various Hilbert subspaces that represent different possible results of the measurement of A.

[36] In Secs. 1.18–1.20 and 1.25, we shall deal with possible objections against this claim, and in Secs. 1.21–1.22 we discuss a general class of cases in which this claim is justified.

[37] See Sec. 2.11.

[38] The state of the ensemble $E_{za}^{\mathbf{OA}}$ of systems **OA** may slightly differ from the state of the ensemble $E_a^{\mathbf{OA}}$ of such object-apparatus systems, as the recoil of **A** may slightly depend upon the preceding state z of **O**. See Wigner,[7] and see sec. 15.2 of the 1971 book of d'Espagnat.[5] Yet, $\psi(E_{za}^{\mathbf{O}}) = \psi(E_a^{\mathbf{O}}) = a$.

of the systems **O** in E_{za} thus becomes irrelevant.) In this case, our *choice*[39] of E_{za} or of E_a instead of E_z may have been influenced by knowledge acquired in an actual individual measurement of A. However, whether our choice is thus motivated or whether it merely is our *whim* to pick from E_z the subensemble E_{za} of cases where measurement of A gives a particular result a, has no influence upon the objective quantum-mechanical probabilities valid in E_z and in E_{za} and in E_a. We conclude that, as long as the rules of the theory are independent of whether or not we are going to make some particular application of the theory, *knowledge* of the result of some measurement should be *irrelevant* for the discussion of these rules.

We therefore do not feel the need, suggested by Wigner,[40, 41] for introducing an influence of consciousness upon quantum-mechanical systems. We are particularly reluctant to accept Wigner's speculations on this matter, where he bases them upon ideas which we reject.[42]

[39] This *choice* is what we call the *reduction* of the state. It is a reduction of the ensemble of which the state is to be discussed.

[40] E. P. Wigner, *Remarks on the Mind-Body Question*, chap. 98 in I. J. Good: *The Scientist Speculates*. (Basic Books, New York, 1962.)

[41] See also E. P. Wigner, *Foundations of Physics*, **1**, 35 (1970). Wigner claims that the "observations" with which quantum theory deals would be *perceptions entering consciousness*, and then invokes the law of action and reaction to argue that, if matter influences consciousness, also consciousness ought to affect matter. As discussed in Sec. 1.21 below, in objective quantum theory we do not deal with individual observations, but with subdivisions of ensembles of observations into subensembles distinguished from each other by different "records" left—possibly on inanimate objects—after identical measurement manipulations. Thus, objective quantum theory does not deal with realistic perceptions, but, like any theory, deals merely with abstractions, and is useful only because one can (by repeating a measurement very often) create circumstances that simulate the idealized premises of the theory, and one finds that the predictions of the theory for the idealized case fairly well describe the statistical results for these simulations of ensembles. Because of the abstract nature of the objective interpretation of quantum theory, Wigner may well be justified in calling the epistemology of quantum mechanics "unrealistically schematized and barren" (*loc. cit.*, top of page 39), and his eloquent plea for a different theory that comes to grips with the sensations of conscious beings has a certain appeal. We agree with Wigner's final remark (on page 44) that at the present not much usefulness of such considerations can be seen, and this is our reason for leaving these *per se* interesting ideas out of the picture of quantum theory which we are presenting here.

He does so where, in the story about his famous "friend," he ascribes a state vector to "friend and object" in a discussion of a measurement performed only once.[43] Shimony, after an attempt at justifying such speculations, comes to the conclusion that they are "counterintuitive in the extreme."[44]

1.12. Schrödinger's cat

We have kept repeating that state vectors describe *ensembles*. How ridiculous it is to assume that a state vector would describe a probability distribution for a single system is well illustrated by the famous example of "Schrödinger's cat."[45] For a discussion, see footnote 11 on page 369 of Ballentine's paper.[13]

Nevertheless, this misunderstanding of ψ frequently creeps up in the literature. For instance, the spirit of Schrödinger's cat seems to be revived where Wigner, talking about his famous friend, writes[46]: "If the atom is replaced by a conscious being the wave function ... (which also follows from the linearity of the equations) appears absurd because it implies that my friend was in a state of suspended animation before he answered my question."[47]

Another example, related to the previous one, is more subtle, and has given rise to considerable misunderstanding. In 1963, Wigner[48] wrote: "... *measurements which leave the system object-plus-apparatus in one of the states with a definite position of the pointer cannot be described by the linear laws of quantum mechanics.*" In 1966 d'Espagnat[49] wrote an interesting paper in which he generalized Wigner's argu-

[42] We refer here to an interpretation of the meaning of the state vector which disagrees with the ideas expressed in Sec. 1.3. For a discussion, see Sec. 1.12.

[43] Reference 40, page 289, and discussion on pages 292–294.

[44] Abner Shimony, *Am. J. Phys.* **31,** 755 (1963). See its page 772.

[45] E. Schrödinger, *Naturwissenschaften*, **23,** 807 (1935), especially page 812. (See also our Preface.)

[46] Reference 40, bottom of page 293.

[47] The "suspended animation" about which Wigner is talking would consist in his friend's not definitely knowing the result of a measurement which his friend had already completed. (Compare our Preface!)

[48] E. P. Wigner, *Am. J. Phys.* **31,** 6 (1963).

[49] B. d'Espagnat, *Supplemento al Nuovo Cimento*, **4,** 828 (1966).

ment. His reasoning started out with the mathematical model which we shall use in Sec. 1.17, which leads after a measurement to the density matrix Eq. (9), or to a linear combination of such density matrices if the initial state was already a mixed state. On the basis of a few assumptions that seemed plausible to him,[50] d'Espagnat then showed that necessarily this density matrix (9) would differ from density matrices of the kind illustrated by our Eq. (8) which appears to be required for validity of the rule **F5**. He therefore concluded that the application of rule **F5** would require an alteration of the density matrix. Using an assumption made in his Introduction, that in a certain sense the possible observables would form a *complete set*, he then showed that such a change of density matrix would not be possible without unjustifiably altering the theoretical predictions for ensembles of subsequent measurements of some of these observables. He concluded that in that case rule **F5** had to be faulty, and that, *if the observables form such a complete set* (as discussed in his Appendix I), "the statement that the pointer takes a definite position is not compatible with the general laws of quantum mechanics."

The validity of this possibly queer-sounding[51] conclusion is not as

[50] The proof given by d'Espagnat hinges on several assumptions. (1) At the top of page 833 of his paper, he assumes that the time dependence of the entire ensemble of systems **OA** shall be given by a *unitary* transformation of the density matrix, as if it were given by one single well-determined hamiltonian. In the laboratory, this is seldom rigorously true. (2) His proof also uses the fact that, though in general a density matrix does not uniquely determine the wave functions over which it describes a probability distribution, different sets of wave functions that could be used are always linear combinations of each other. In particular, the wave functions which by rule **F5** should form a mixed state are linear combinations of the wave functions which are mixed at the end of the measurement. We find no fault with this conclusion, though d'Espagnat (in his Appendix II where this is discussed) considers again only unitary transformations from one complete orthonormal set of wave functions to another (where he identifies the probabilities with the eigenvalues of the density matrix), so that he underestimates the freedom of choice of interpretation of a density matrix. (For instance, in our own footnote 1 we have shown that the probabilities of the wave functions mixed need not be equal at all to these eigenvalues.)

[51] It sounds queer to those who, in the conventional way, count rule **F5** *among* the general laws of quantum mechanics. In d'Espagnat's context, however, by "general laws" merely the time dependence by a Schrödinger equation is meant.

general as d'Espagnat's statement might make us believe. When measurements are made in the laboratory, the "mixed state" of the object-plus-apparatus system usually is characterized not merely by the initial state being a probability distribution over pure states, but also by a dependence of the governing hamiltonian, or at least of its time integral over the duration of the measurement, on statistically distributed parameters. Thus, the effect of the time dependence of the density matrix is no longer a unitary transformation, and a pure state may easily change into a mixed state. As an example will easily show (see Appendix D), this may often lead to a final mixed state with a density matrix of the kind required by rule **F5**.

If, however, the time integral of the hamiltonian has no uncertainties, d'Espagnat's conclusion is valid. The conclusion to be drawn from this *within the limitations of objective quantum theory* (of which the assumption of validity of the rule **F5** for successful measurements is an integral part) should then be that apparently after a successful measurement the observables which are to be considered measurable *can no longer form a complete set*. This is exactly what we shall find in Sec. 1.21.

Others, however, have regarded the result of Wigner and d'Espagnat (that the incident state could not by the Schrödinger equation evolve into a mixed state interpretable as a probability distribution over definite measurement results) as if it were unconditionally valid, forgetting about d'Espagnat's condition that the observables would form a complete set. They have concluded that according to quantum theory the pointer position of the measuring apparatus at the end of a measurement would still be indeterminate. Fine[52] has stressed this point of view in an elaboration of d'Espagnat's work. Wigner has taken this point as evidence for the need for assumption of an inter-

[52] A. Fine, *Phys. Rev.* D, **2,** 2783 (1970). See on page 2787 in his **DISCUSSION.** For comments, see Moldauer.[53] Unfortunately, in a recent paper, Maxwell[54] has taken Fine's claim to be valid.

[53] P. A. Moldauer, *Phys. Rev.* D, **5,** 1028 (1972).

[54] N. Maxwell, *Am. J. Phys.* **40,** 1431 (1972). His views are in conflict with the fundamentals of objective quantum theory discussed in our Secs. 1.3–1.7.

action between our own consciousness and the outside world: The pointer position supposedly would remain indeterminate until our consciousness finds it to be in one position or in another (never in several positions simultaneously).[55]

The statement that the pointer position in a single measurement would remain objectively indeterminate until someone becomes conscious of the result of the measurement may fit into a solipsistic picture of nature,[56] but it does not fit into a picture that ascribes to outside nature any objective realism.[44] In as far as physics tries to give an objective description of this outside world, therefore, it either ignores or rejects this point of view. It is not our task here to argue whether or not physics is right in this attitude. Objective quantum theory simply states as a fact that observables can be successfully measured upon single systems, which means that at the end of the measurement some record will have been created of its definite result (known or unknown to any living mind). This objective point of view agrees with what our mind experiences when it becomes conscious of a result. It is admitted that objective (outside) reality may influence the mind, but the "reaction" of the mind upon the outside world is in physics commonly *neglected*, particularly as it clashes with any ideas of causality to assume that the mind would influence the realization of a definite result of measurement, as the mind usually interacts with the results only after the measurements have been completed.

Returning to our starting point and summarizing, we see that d'Espagnat's conclusion about indefiniteness of the pointer position, when d'Espagnat's conditions for its validity are forgotten, revives the paradox of Schrödinger's cat. That is, it then would contradict rule **F5** of objective quantum theory, by claiming that the result of a measurement (like whether Schrödinger's cat is dead or alive) would not only be unpredictable, but even would remain indefinite *after* some intrument would have recorded the absence or presence of the cat's heart beat. This claim of indefiniteness, when made, bases itself upon a tacit assumption that a pure state ψ that would be a

[55] Private conversations. See also references 40, 44, 56, and 41.
[56] E. P. Wigner, *The Monist*, **48**, 248 (1964).

superposition of states in which the cat would be dead and in which it would be alive would mean for a single cat a state of fractional truths in both claims that this cat would be dead or alive. This assumption contradicts the meaning in objective quantum theory given to such a superposition ψ by rule **F5**, according to which ψ merely determines a frequency distribution in an ensemble, while for each single cat life or death would after its measurement be definitely established objectively, even though unpredictable beforehand.

1.13. Is reduction of the state vector after a measurement either mandatory or illogical?

From Sec. 1.11 it should be clear that the answers to both questions are "No." That the same results derivable with reduction of the state vector are (in a more clumsy way) also obtainable without it, was shown already in 1957 by Everett.[10] It also follows from the use of the word "choice" in the definition of "reduction,"[39] that reduction cannot be mandatory.

On the other hand, in a theory based upon rule **F5** quoted on page 16, the *possibility* of reduction of E_z to one of its subensembles E_{za} determined by the results of measurements of A follows directly from the wording of rule **F5**, which postulates the *existence* of these subensembles in the ensemble of objects **O** after the measurement.

Therefore, reduction is merely a *privilege* that we may use to our advantage.[36]

Returning to the question whether reduction could be mandatory, some authors claim that the appearance of tracks in a Wilson chamber would prove that reduction of states by interactions would be unavoidable. Once the first two droplets have been created, the next droplet must lie in approximately the same direction. This is supposed to prove that "the" state of propagation of the particle must have been reduced from what may have started as a spherical wave, to the state of a particle traveling in a particular direction.

The argument obviously is fallacious. All the third droplet proves is a correlation between the directions of the vectors from the first droplet to the second, and from the second to the third one, as required (for weak perturbation of the particle upon drop formation) by the conservation laws.[57] Reduction after observation of the second droplet provides the easiest way of describing this correlation, as it splits up the original possibly spherical ensemble around the first droplet into subensembles in each of which the first vector is fixed; then, of course, the second vector has a tendency of lying in about the same direction. But, we *may* refuse to reduce the ensemble, and keep studying the original garbled ensemble in which the first vector had all possible directions. Then, also the second vector has all possible directions. In this garbled ensemble, the correlation between the two vectors would still exist. It only would require more complicated formulas to describe it,[58] and therefore we usually prefer to make use of our privilege and perform a reduction whenever allowed to do so.

1.14. Does reduction change the state of any well-defined entity?

The answer to this question is again: No. Sticking to the terminology of Sec. 1.8 (and remembering Schrödinger's cat), we do not assign state vectors (or probability distributions over state vectors) to individual objects **O**. We assign them only to *ensembles*. When we simplify the procedure of successive applications of quantum theory to successive measurements, by introducing a new (reduced) ensemble E_{za} (or the predictively equivalent ensemble E_a) for predicting the future statistical behavior of those members of an original ensemble E_z on which by an ideal measurement of A some selected result (corresponding to an eigenfunction) a has been found, this selection[39] of subensemble E_{za} from E_z *does not change the state vector or density matrix* of either E_z or E_{za}. Merely, the newly chosen ensemble has a state of its own, and quantum theory tells us objectively what it is.

[57] Compare also Appendix A.

[58] See Appendix E on how this correlation could be described wave-mechanically.

1.15. The time at which a state is reduced by a measurement

From Sec. 1.13 it is clear that asking for the time at which reduction takes place is a question without a scientific answer, as there are no rules about when one should make a choice. One might never make it.

The literature, however, contains a variety of speculations about the time of reduction,[59] as if the question made any sense. Usually, such discussions show the misunderstanding that the state would be a property of a single system.

Additional confusion is caused by the fact that by "reduction" not all authors understand the same stage in the evolution from the initial state of the ensemble before the measurement, to the final state of the reduced ensemble. In the initial ensemble, firstly interactions take place between object and apparatus. More often than not,[60] during this time the interaction washes out any phase relations that before the measurement possibly existed between the components of the initial state that give rise to probabilities for different outcomes of the measurement.[61] (Such a loss of coherence presumes, of course, that the state of the object-plus-apparatus system will be a mixture with regard to the values of some parameters describing microscopic details of the meeting between object and apparatus, and that these parameters will affect the time integral of the hamiltonian of the Schrödinger equation for this system, as illustrated in Appendix D, so that uncertainties in phase differences that initially were negligible become during the interaction and only for different results of the measurement magnified to values that modulo 2π are randomly distributed.) This is one way[61] in which a state which origi-

[59] Examples are the story about Wigner's friend,[40] and the fantasy which d'Espagnat erroneously ascribes to Bohr, that the reduction would "occur" already before the measurement is made, at the time the measurement is planned. (See Sec. 2.13 and Appendix C.)

[60] I purposely do not say "always."

[61] This loss of coherence (discussed, for instance, by D. Bohm on pages 600–608 of his *Quantum Theory* [Prentice Hall, New York, 1951]) is *not a necessary* condition for arriving at our final conclusions. In Secs. 1.19 through 1.23, we will discuss other reasons why the final state of the initial ensemble after a successful measurement may for all practical purposes be regarded as a mixed state.[62] When coherence *is* lost, however, this is *sufficient* for causing a mixed final state.

nally was hard to distinguish from a pure state might develop clearly the characteristics of a mixture of possible outcomes of the measurement, as all effects of interference between different outcomes (characterized by off-diagonal matrix elements of the density operator between different measurement results) would appear multiplied by arbitrarily distributed sines and cosines that in the ensemble would average to zero.[62]

After thus or otherwise the ensemble has reached this stage, we call it *reducible*. That is, for all practical purposes it has become a mixture (= classical combination) of subensembles of which each corresponds to a single measurement result; so, if one result ($A = A_N$) is of particular interest to us (like when we want to select the cases where it is realized experimentally), we may "reduce" our initial ensemble of all measurements, to the particular subensemble E_N corresponding to this result.

Finally, from the state vector $\psi_N^{\mathbf{OA}}$ for the pure state of this subensemble $E_N^{\mathbf{OA}}$ of $\mathbf{OA}$ systems, we may in case of an *ideal* measurement omit the factor $\psi_N^{\mathbf{A}}$ describing the state $A = A_N$ of $\mathbf{A}$, and obtain the state vector $\psi_N^{\mathbf{O}}$ for the final reduced state of the ensemble $E_N^{\mathbf{O}}$ of objects $\mathbf{O}$ prepared by the measurement. This $\psi_N^{\mathbf{O}}$ then may be used as an initial state in the description of subsequent measurements made upon this ensemble of objects $\mathbf{O}$ alone.

In von Neumann's projection postulate, all these steps are taken at once. They are, however, separate steps in the explanation of what is taking place, and different authors have applied the name "reduction" to different steps in this sequence.

To us, reduction is *not* the change from a pure initial state to a mixed state. Rather, it is the transition by choice, from this (reducible) mixed state, to a pure state corresponding to a selected result of measurement.[39] While the preceding stages (including a possible loss of

[62] A successful measurement leads to states such that all operators corresponding to what a *physicist* would consider to be observables will have vanishing matrix elements between any pair of these states corresponding to two different [72] results of that successful measurement. See Sec. 1.21.

coherence) may be governed by the Schrödinger equation, the selection of a subensemble which *we* call the reduction is *not* governed by it.

No reduction should be performed unless the state of the original ensemble first has become reducible. As we will find in Secs. 1.19–1.21, this state will automatically become reducible after a measurement has been *successfully* completed. This completion includes the recording of the result for later reference, which in the theory is idealized as the entering of an indelible record into some register that we will call the "memory."

1.16. Common objections against the above conclusions

Those who have any objections against the preceding sections and claims contained therein, might well read chap. 4 of a paper by Ballentine,[13] in which he shows how most of such objections are based upon the assumption that state vectors would describe individual systems. The latter assumption, when made, often is based upon wishful thinking only. (Compare Sec. 1.4 above.) It should be clear by now that this assumption cannot be accepted in objective quantum theory.

In the following sections, we shall give additional justification of our various claims.

1.17. Conventional formalism of the theory of measurements

In Sec. 1.11 we considered measurement of A on systems prepared by preceding ideal measurements of Z, choosing as population of our ensemble E_z the sample in which this preceding measurement yielded the particular result z. In order to avoid confusion with what Moldauer[53] calls "measurements of type Z,"[75] from here on throughout Chapter 1 I will write 0 (zero) instead of z, so that A is measured by apparatus **A** on the ensemble $E_0^{\mathbf{O}}$ (with state vector $\psi_0^{\mathbf{O}}$) of systems **O**. Let ϕ_n be the eigenfunctions of $\mathcal{A}$ to the eigenvalues A_n (here for simplicity treated as if they were discrete and nondegenerate), and let

$$\psi_0^{\mathbf{O}} = \sum c_n \phi_n . \tag{1}$$

Let ξ_0 be the initial state of the ensemble of apparatus used.[63] Then, the initial state of the ensemble $E_{(1)}^{\mathbf{OA}}$ of systems **OA** is[64]

$$\psi_{(1)}^{\mathbf{OA}} = \xi_0 \sum c_n \phi_n . \tag{2}$$

During the measurement, the apparatus is supposed to take on some "pointer position" state ξ_n. Therefore it is conventional to postulate that the Schrödinger equation shall cause a unitary transformation from (2) to[64]

$$\psi_{(2)}^{\mathbf{OA}} \approx \sum c_n \xi_n \tilde{\phi}_n , \tag{3}$$

where we may expand the perturbed object state by

$$\tilde{\phi}_n = \sum g_{nm} \phi_m \qquad \left(\text{with} \quad \sum_m |g_{nm}|^2 = 1\right) . \tag{4}$$

If the measurement is an ideal one, we have

$$g_{nm} = \delta_{nm} . \tag{5}$$

Wigner[7] has shown that one cannot in general assume Eq. (3) with or without (5) to be rigorously valid if the apparatus is not

[63] We treat this here as if it were a pure state. Almost certainly, the ensemble of apparatus will be in a mixed state. Then, there are many different $\xi_{0(j)}$ with probabilities $p_{(j)}$, $j = 1, 2, 3, \ldots$. Correspondingly, there will be probabilities $p_{(j)}$ for the results obtained by inserting $\xi_{0(j)}$ for ξ_0. Note that for nonideal measurements the g_{nm} in Eq. (4), and correspondingly the $\tilde{\phi}_n$, may somewhat depend upon which $\xi_{0(j)}$ is used. With sufficient generalization of this kind, one may regard Eq. (3) as describing at least subensembles of $E_{(2)}^{\mathbf{OA}}$, even if **A** is a macroscopic apparatus.

[64] Subscripts (1) on E, ψ, etc. indicate consideration of the ensemble $E^{\mathbf{OA}}$ of systems **OA** in which the **O** are taken from $E_z^{\mathbf{O}} = E_0^{\mathbf{O}}$ and the **A** are taken from an ensemble $E_0^{\mathbf{A}}$ in the state ξ_0. Similarly, subscripts (2) refer to this same ensemble $E_{(1)}^{\mathbf{OA}}$ *after* **O** and **A** have interacted.

infinitely large.[65] (Even if the apparatus were infinitely large, the above formulas should be understood as describing the measurement merely schematically.) In particular, when there is in **OA** conservation of an observable not commutative with A, no unitary transformation would lead from (2) to (3). It is possible, however, to maintain the unitarity of the transition $\psi_{(1)} \to \psi_{(2)}$ for finite apparatus by adding to (3) some terms that may be made small by increasing the size of the apparatus **A**.[65] In the following, we shall neglect such terms.

Consider the subensemble of measurements in which the pointer reading has $n = N$. The state vector $\psi_N^{\mathbf{OA}}$ of this subensemble $E_N^{\mathbf{OA}}$, according to rule **F5** of Sec. 1.11, is then given by[66]

$$\psi_N^{\mathbf{OA}} = \xi_N \, \tilde{\phi}_N, \tag{6}$$

and the weight of $E_N^{\mathbf{OA}}$ inside $E_{(2)}^{\mathbf{OA}}$ is given by[67]

$$w_N = |c_N|^2. \tag{7}$$

Therefore, this axiom **F5** is equivalent to postulating that, after an "always successful"[68] measurement by **A** on **O**, the state of the ensemble $E^{\mathbf{OA}}$ shall become describable as a *mixed state* with density operator

$$W^{\mathbf{OA}}_{\text{reducible}} = \sum_n |c_n|^2 \, |\xi_n \, \tilde{\phi}_n\rangle \langle \xi_n \, \tilde{\phi}_n| \, . \tag{8}$$

This state (8) we called *reducible* (p. 25), but is not yet *reduced*.[39]

According to Daneri *et al.*,[69] ergodic processes may "magnify"

[65] See Appendix F.

[66] Obviously the rule **F5** presupposes that we may neglect the probability that the outcome of the measurement is not recognizable as any definite value for A (compare Appendix F), since otherwise $\sum w_N < 1$, while $\sum |c_N|^2 = 1$.

[67] The probability for the pointer position ξ_N in $E_{(2)}^{\mathbf{OA}}$ is $\int dq_{\mathbf{O}} \, |\int dq_{\mathbf{A}} \, \xi_N^* \, \psi_{(2)}^{\mathbf{OA}}|^2 = \int |c_N \, \tilde{\phi}_N|^2 \, dq_{\mathbf{O}} = |c_N|^2 \sum_m \sum_{m'} g^*_{Nm} \, g_{Nm'} \, \langle \phi_m | \phi_{m'} \rangle = |c_N|^2 \sum_m |g_{Nm}|^2 = |c_N|^2 = w_N$. Using these probabilities, we find $\langle A \rangle = \sum w_N A_N = \langle \psi_0^{\mathbf{O}} \, |A| \, \psi_0^{\mathbf{O}} \rangle = \langle \psi_{(1)}^{\mathbf{OA}} \, |A| \, \psi_{(1)}^{\mathbf{OA}} \rangle =$ the expected mean of A in the ensemble $E_0^{\mathbf{O}}$ or $E_{(1)}^{\mathbf{OA}}$. Note that in general $\langle \psi_{(2)}^{\mathbf{OA}} \, |A| \, \psi_{(2)}^{\mathbf{OA}} \rangle$ will differ from this.

[68] That is, a measurement by an "infinitely large" apparatus, described schematically by (3) *without* addition of terms suggested by Wigner[7] (see Appendix F) wich would describe final apparatus states that cannot be interpreted as any pointer position at all.

the pointer position to something macroscopically conspicuous. This idea may be incorporated in the above formalism in either of two ways: (I) One might consider the macroscopic instruments of Daneri *et al.* to be part of **A**, and regard ξ_N as formally already describing the final macroscopic pointer position. (II) Better, we may incorporate the macroscopic parts of the apparatus in the "memory" **M** discussed in Sec. 1.21 below, and the transition to the final state of the memory includes the ergodic processes.

Properly speaking, ergodic processes like many other parts of the measurement process can, of course, be described by Schrödinger equations only in the sense of footnote 63, with different component pure states in the over-all mixed state being entitled to their own hamiltonians, so that the density matrix in general does not evolve by unitary transformation. (Compare footnote 50 on page 19.) This makes the entire conventional formalism of the theory of measurements an oversimplified idealization. Nevertheless, it has its value as a starting point from which a more realistic theory can proceed, by adding details in the spirit of footnote 63.

The so-called "quantum-mechanical measurement problem" is whether the transition from (3) to (8) is justifiable. The density operator describing (3) directly is

$$W_{(2)}^{\mathbf{OA}} = \sum_{n}\sum_{n'} c_n^* c_{n'} \, |\xi_{n'}\phi_{n'}\rangle \langle \xi_n \phi_n| , \tag{9}$$

and differs from (8) by the terms with $n' \neq n$.

D'Espagnat[49] and Fine[52] have reasoned that the state described by (3) and (9) has more credibility than (8) because (3) is a direct consequence of the linear Schrödinger equation alone, and they have claimed that a final state (3) would mean that after the measurement

[69] A. Daneri, A. Loinger, and G. M. Prosperi, *Nuclear Phys.* **33,** 297 (1962) and *Il Nuovo Cimento*, **44B,** 119 (1966). A clear summary of this work is found in a paper by Rosenfeld.[70]

[70] L. Rosenfeld, *Progress Theoret. Phys.* Supplement, Extra Number on Meson Theory, 222 (1965).

the pointer would not show any definite position. (Specters of Schrödinger's cat.[71]) Using the fact that the $\xi_n\tilde{\phi}_n$ form an orthogonal[72] set of eigenfunctions of the pointer position observable, Moldauer[53] has answered that measurement of the pointer position[73] on the ensemble $E^{\mathbf{OA}}$, according to the interpretation **F5** of the state vector (3), will give the same result (a probability distribution over unambiguous pointer positions) when $E^{\mathbf{OA}}$ is in the pure state (3) or (9), as when it is in the mixed state (8).

This conclusion is correct if one regards rule **F5** as an axiom, or at least as a valid and indispensable rule needed for the probability interpretation of the quantum-theoretical formalism. (This is the point of view which we have taken in the *preceding* sections.) However, as a justification of **F5** itself, it is not very convincing, being from that point of view a circular reasoning, showing at most a certain selfconsistency of this rule.

1.18. What could be wrong with the probability interpretation of the state vector?

Some people are not willing to accept **F5** as one of the axioms of quantum theory, because they feel that its assumption might be incompatible with other rules of the theory.[71] As nevertheless they find rule **F5** successfully used in many applications of the theory, they would be willing to replace the axiom **F5** by some theorem, derivable from the other axioms of the theory, and approximately equivalent to **F5** for all practical purposes, but having a conditional validity.

Others claim that they can apply quantum theory without invoking the reduction formalism at all. These people replace *probabilities in*

[71] See Sec. 1.12.

[72] We use here the assumption that all measurement results are distinctly recognizable. (Compare footnote 68.)

[73] Either *measurement* or *registration* of this position is required for successfully completing the measurement of A.

prepared states by *relative probabilities*,[74] and use a formula for the latter which is derivable from the rule **F5**, so that a justification of **F5** remains an important task, even if we do not make the change of ensemble which we called "reduction."

Let us here briefly summarize the two main arguments against taking **F5** simply as an axiom.

In the first place, we have seen above that the validity of **F5** is limited. It may be valid for idealized measurements by instruments that can absorb certain types of recoil without being noticeably affected, but it certainly cannot be valid without qualification if there is a nonvanishing probability that the measurement of A will not lead to any definite value of A at all.[65, 66]

In the second place, suppose that after the measurement of A on $\mathbf{O}$ we measure on $\mathbf{OA}$ a new observable B for which the matrix elements $\langle \xi_n \tilde{\phi}_n | B | \xi_{n'} \tilde{\phi}_{n'} \rangle$ for $n \neq n'$ do not all vanish.[75] Then, if $W_{(2)}^{\mathbf{OA}}$ of Eq. (9) is the correct density operator for the ensemble $E_{(2)}^{\mathbf{OA}}$ at the end of the first measurement, use of $W_{\text{reducible}}^{\mathbf{OA}}$ of Eq. (8) is going to give wrong results by suppression of important interference contributions to the mean value of B given by

$$\langle B \rangle = \text{Trace}\,(W^{\mathbf{OA}}\, B). \tag{10}$$

[74] Let w_N be the probability of occurrence of the pointer position ξ_N in the ensemble $E_{(2)}^{\mathbf{OA}}$ in the pure state $\psi_{(2)}^{\mathbf{OA}}$ of Eq. (3). Let a measurement B be made[68] upon the systems $\mathbf{O}$ in this ensemble. If ζ_m are the pointer position states of **B**, if $\mathcal{B}\chi_m = B_m \chi_m$ and $\tilde{\phi}_n = \sum_m d_{nm} \chi_m$, and if $\zeta_0 \chi_m \rightarrow \zeta_m \tilde{\chi}_m$ in apparatus **B**, the ensemble $E_{(2)}^{\mathbf{OAB}}$ goes into a state $\psi_{(3)}^{\mathbf{OAB}} = \sum_n \sum_m c_n d_{nm} \xi_n \zeta_m \tilde{\chi}_m$, and $w_{Nm} = |c_N|^2\,|d_{Nm}|^2$ is the probability for having in $E_{(3)}^{\mathbf{OAB}}$ simultaneously pointer positions ξ_N and ζ_m. Without explicitly reducing $\psi_{(2)}^{\mathbf{OA}}$ first to $\psi_N^{\mathbf{OA}}$ of Eq. (6), we may then calculate the relative probability $w_{(N)m}$ of ζ_m for given ξ_N, by $w_{(N)m} = w_{Nm}/w_N$ with $w_N = \sum_m w_{Nm}$. See, for instance, Moldauer[53] below his eq. (12). The use of **F5** here made for obtaining w_{Nm}, however, tacitly implies reducibility of the state of $E_{(3)}^{\mathbf{OAB}}$.

[75] Measurement of such a B is what Moldauer[53] calls a "measurement of type Z." When such measurements are possible, the state of the ensemble $E_{(2)}^{\mathbf{OA}}$ with the state vector of Eq. (3) and the density operator (9) determines, for the ensemble $E_{(2)}^{\mathbf{O}}$ of the objects $\mathbf{O}$ that are part of the $\mathbf{OA}$ in $E_{(2)}^{\mathbf{OA}}$, a state that is rather similar to a mixture and that d'Espagnat calls an *improper* mixture, because it is distinguishable from *true* mixtures by measurements of B. (See reference 4, pages 36–41, or reference 5, sec. 6.3.)

These arguments against *axiomatization* of **F5**, and a method of finding a *qualified* justification of **F5**, are well illustrated in the following discussion of an example considered already by Bohr and Einstein in their conversations on quantum theory.[76]

1.19. The example of Einstein and Bohr

A beam of photons[78] hits a fixed vertical screen with a single horizontal slit. The photons diffracted through this slit hit a second screen with two slits parallel to the first one. The photons diffracted through the latter two slits hit a vertical photographic film.[79] Interference fringes are formed upon the film.

We know that photons traverse the screen only at the slits. (This may be verified by placing photon detectors close behind the screens.) It therefore seems natural to ask through which slit the photons traveled which hit the film at certain places in the interference fringe pattern.

One way to answer such a question is by making up some theoretical rule. For instance, a theory of Bohm,[80] which to some extent may be regarded as a hidden-variables theory of the first kind,[81] gives a rule of this sort for electrons[78] forming an interference pattern. However, there is no way of experimentally verifying this rule, as Bohm's theory agrees with quantum theory in predicting that an *experimental* determination of the slits through which the individual particles travel will automatically wash out the interference pattern.

[76] Niels Bohr, *Discussion with Einstein on Epistemological Problems in Atomic Physics*, chap. 7 in reference 77. See, in particular, pages 215–220.

[77] P. A. Schilpp, *Albert Einstein: Philosopher-Scientist*, vol. 7 of The Library of Living Philosophers. (Evanston, Illinois, 1949. Republished by Open Court, La Salle, Illinois, 1970.)

[78] Other particles could be substituted for the photons. This may require replacing the screens with slits by different devices which stop and diffract the beams.

[79] See reference 76, page 216, fig. 3.

[80] D. Bohm, *Phys. Rev.* **85,** 166 and 180 (1952).

[81] See reference 3, Part Two.

Einstein suggested to Bohr a method of determining the slit through which a photon would reach the film, without in any way perturbing the photon waves by gadgets at the slits.[82] For this purpose, the first (fixed) screen with single slit was to be replaced by a light-weight screen-with-slit suspended on springs.[83] Whenever a photon passing through this slit would be diffracted slant up toward the upper slit in the second screen, the first screen would recoil downward because of conservation of momentum. The photons toward the lower slit would be recognizable by a corresponding upward recoil of the first screen with slit.

During the discussion with Bohr, it was soon found that also this determination of the slit through which the photon would reach the film would wash out the interference fringes: The accuracy in the determination of the momentum of the first screen needed for determining toward which slit the photon is traveling, suffices by Heisenberg's indeterminacy relation for making the vertical location of the first slit sufficiently uncertain that the location of the fringes on the film is no longer determined within a latitude smaller than the distances between the fringes.

Let us here formulate this problem mathematically in a form reminiscent of our formulas of Sec. 1.17. We shall use a notation as if we are talking about electrons rather than about photons.[78]

Let ϕ_1 and ϕ_2 be the (individually normalized) parts of the wave reaching the film through the upper and through the lower slit, respectively. If the arrangement is not entirely symmetric, the coefficients c_1 and c_2 in the expansion

$$\psi_0^{\circ} = c_1\phi_1 + c_2\phi_2 = \sum c_n \phi_n \tag{1a}$$

of the total wave function need not have equal absolute values.

If the Einstein alteration of the first screen **A** is made, let ξ_0 be the initial state of this screen, and let ξ_1 and ξ_2 be its states of downward and of upward recoil which accompany the states ϕ_1 and ϕ_2 of the

[82] See reference 76, page 217.
[83] *Ibidem*, page 220, fig. 5.

particles **O** traveling *via* the upper and the lower slit. In an ensemble of measurements, as the particles **O** travel through the slit **A**, the state of the ensemble $E^{\mathbf{OA}}$ changes from $\psi_{(1)}^{\mathbf{OA}} = \xi_0 \psi_0^{\mathbf{O}}$ into

$$\psi_{(2)}^{\mathbf{OA}} = \sum_n c_n \xi_n \phi_n, \tag{3a}$$

where we have assumed that the change $\phi_n \to \tilde{\phi}_n$ of the particle waves by the recoil of the first screen was negligible.

It was just this change from ϕ_n to $\tilde{\phi}_n$ that was invoked by Bohr for explaining why the interference pattern on the film would be washed out. However, *even without* this change of ϕ_n by the change of location of the diffraction center at the first screen, we will find the interference pattern on the film to be washed out.

The reason is that the particle density on the film is obtained from $|\psi_{(2)}^{\mathbf{OA}}|^2$ by integration over the "coordinates" $q_{\mathbf{A}}$ of system **A**. Since, for *distinguishable* recoils of **A**,[72]

$$\int \xi_n^* \xi_{n'} \, dq_{\mathbf{A}} = \delta_{nn'}, \tag{11}$$

there cannot exist any interference cross terms between different n and n' in the particle density observed on the film.

Thus here, as is usually the case, the *orthogonality of the pointer position states among each other* prevents the occurrence of interference effects, and replacement of the "correct" density operator (9) by the "reducible" density operator (8) will not alter one bit the predictions made for the particle beam after traversing the apparatus **A** (suspended first slit) which measures the observable A (= choice of slit in second screen traversed by the particle) by distinction between ξ_1 and ξ_2.

In a language often found in older discussions of this topic,[84] in which one uses the fact that the different orthogonal ξ_n have undetermined phase factors $\exp(i\,\alpha_n)$ relative to each other, it might also be said that the occurrence of the different ξ_n in the different terms of Eq. (3a) for $\psi_{(2)}^{\mathbf{OA}}$ provides each ϕ_n in the sum with an independent

[84] Compare, for instance, Bohm's book quoted in reference 61.

phase factor.[85] Thus, the ensemble $E^{\mathbf{O}}_{(2)}$ of components $\mathbf{O}$ of systems $\mathbf{OA}$ in $E^{\mathbf{OA}}_{(2)}$ is in a mixed state consisting of a probability distribution over undetermined relative phase angles α_n in

$$\psi^{\mathbf{O}}_{(2)} = \sum c_n \phi_n \exp(i\alpha_n). \tag{12}$$

The density operator of this mixed state is the average over all α_n, of $\psi^{\mathbf{O}}_{(2)}(x)\,\psi^{\mathbf{O}}_{(2)}(x')^* = \sum_n \sum_{n'} c_n c^*_{n'} \phi_n(x)\, \phi^*_{n'}(x') \exp(i\alpha_n - i\alpha_{n'})$, and is the same one as obtained, for $E^{\mathbf{O}}_{(2)}$, from $W_{\text{reducible}}$ [Eq. (8)] if the latter is used for $E^{\mathbf{OA}}_{(2)}$. As all terms with $n \neq n'$ vanish on the average, it corresponds to classical probabilities $|c_n|^2$ for the pure states $\psi^{\mathbf{O}}_n = \phi_n$. This justifies reduction, for each subensemble corresponding to a particular slit recoil ξ_N, to the case of photons traveling through a selected slit N in the second screen. There are in this case no interference effects between states $\xi_n \phi_n$ of the $\mathbf{OA}$ system with different n-values.

We see, however, that the incoherence brought in by the $\exp(i\alpha_n)$ factors is not necessary for arriving at these results. As far as presence or absence of interference fringes on the photographic film is concerned, the mere orthogonality of the ξ_n among each other suffices.

These arguments against interference effects are merely strengthened when with Bohr we replace the ϕ_n in (3a) and (12) by $\tilde{\phi}_n$ that feature an uncertainty in the location of the slit $\mathbf{A}$.

Considering the generality of the argument, one might start to wonder how interference fringes *ever* can be observed behind the second screen. The answer is very simple. In order to get a good interference pattern behind the second screen, we need a *fixed* slit in the first screen. The state ξ_0 of that screen then is not a state of known momentum, but it is a state of given position, and it remains unchanged $= \xi_0$ as the particles pass.[86] Therefore, $\psi^{\mathbf{OA}}_{(2)}$ is in this case not given by (3a), but is unchanged from $\psi^{\mathbf{OA}}_{(1)}$. To say the same differ-

[85] Compare Appendix D, and see footnote 87 on page 36.
[86] As in Wigner's 1952 example,[7, 65] the first screen must be massive enough to make its recoil go unnoticed. For obtaining a predictable interference pattern on the film, also the second screen must be massive.

ently: In (3) or (3a) the different ξ_n are replaced by one common ξ_0. In that case, Eq. (11) is of no help in eliminating interference effects between the contributions to ψ from different states ϕ_n.

Translating this back into the language of Sec. 1.17, we see that interference between the parts of ψ corresponding to different values A_n of the observable measured is possible if the condition of footnote 72 is not met, and the pointer positions ξ_n coincide. That is, interference between different ϕ_n states may become possible when the measurement of A by apparatus **A** *has not succeeded* in distinguishing the ϕ_n states from each other by recognizably different apparatus states ξ_n.

1.20. Wigner's 1963 example

We shall now apply the above to Wigner's 1962 example published in 1963.[48] Here, silver atoms with their spins initially polarized in the $+x$-direction travel in the y-direction first through a Stern–Gerlach apparatus **A** which measures their spin in the z-direction. Behind **A**, magnetic fields bend the two beams together. After the beams have been reunited, it is assumed that, by careful regulation of the path difference along the two coherent beams in **A**, the recombined wave of particles **O** shall again describe atoms with their spin in the $+x$-direction, as it is hoped that the two beams by this time have not yet lost their coherence by the phase difference caused between the two beams by the term $-\boldsymbol{\mu}\cdot\mathbf{B}$ in the hamiltonian, a term which differs for $S_z = +\hbar/2$ from what it is for $S_z = -\hbar/2$.[87]

Next, an apparatus **B** measures the x-component of the spin. If Wigner's hopes about coherence in this imagined experiment are fulfilled, he correctly claims that it would be wrong to replace the

[87] After traveling a distance y in a field $\mathbf{B} \simeq 1000$ gauss at speed $v \simeq 150$ m/sec, this phase difference would be[61] $\simeq 2\mu_B |\mathbf{B}| y/v\hbar \simeq 10^6\, y$/cm, which fluctuates many times 2π if y fluctuates one micron, and which fluctuates even more as thermal motion makes v uncertain. In Wigner's experiment, this washing out of coherence is assumed to be somehow prevented.

density operator (9) at the end of apparatus **A** by the reducible density operator (8).[88]

In this case, the ξ_n are the spatial wave functions of the silver atoms describing whether in **A** they are in the upper or in the lower beam, while correspondingly the ϕ_n are their spin functions for $S_z = +\hbar/2$ and for $S_z = -\hbar/2$. The magnetic field behind **A** now causes a transition $\psi_{(2)}^{\mathbf{OA}} \rightarrow \psi_{(3)}^{\mathbf{OA}}$ to

$$\psi_{(3)}^{\mathbf{OA}} = \xi_{(3)} \sum c_n \phi_n, \tag{13}$$

where $\xi_{(3)}$ is the reunited beam. This is comparable with the situation discussed in the last part of the preceding section, where the pointer positions coincided. That is, in (13) we have succeeded in making an *unsuccessful* measurement by **A**, and therefore interference effects are possible, and the state is not reducible. Nobody (I hope) has ever claimed that reduction should be performed after an unsuccessful measurement.

1.21. Everett's memory state vector

The above situation immediately changes when we insert any apparatus by which the presence or absence of atoms in the beams is registered before the beams are bent together. In the old language of undetermined phases, this inserts factors $\exp(i\,\beta_n)$ into the terms with ξ_n.

Many people consider that language too oversimplifying. Just as it is nicer to introduce pointer position state vectors ξ_n as in Eq. (3), than to use phase factors as in Eq. (12), it is nicer to introduce state vectors μ_n of the registration apparatus, than to introduce phase factors $\exp(i\,\beta_n)$.

The registration apparatus takes the place of von Neumann's "observer of the observer." One could, of course, complicate matters

[88] The reducible (mixed-state) density operator in this case would yield a 50–50 prediction of probabilities for either sign of S_x, instead of the desired 100% certainty of $S_x = +\hbar/2$.

by introducing an infinite sequence of observers, as is sometimes suggested. Instead, we will below introduce a very simple postulate which cuts off this sequence right behind the recording gadget or "memory" **M**.[89]

As it does not matter very much how the recording of the outcome of a measurement is performed, as long as it somehow is done successfully, I feel free to schematize and idealize it. A good schematization is found in the much-spurned 1957 paper by Everett,[10] which makes interesting reading even to those who do not concur with his final conclusions.

Consider an ensemble of measurements by an ensemble of apparatus on an ensemble of objects with the records stored in an ensemble of memories **M**. We describe the state of the latter ensemble $E^{\mathbf{M}}$ schematically as a pure quantum-mechanical state, even though the memory probably would be macroscopic and in a state ill-defined at the atomistic level. The interpretation of such a schematization is similar to the interpretation of a description of a pointer position by a quantum-mechanical state ξ_n. [See footnote 63 on page 27, and the text between Eqs. (8) and (9).] One advantage of using a quantum-mechanical description is that it shows that under certain circumstances a quantum-mechanical system might do either job, as in the case of Wigner's 1963 example, where ξ_n is the spatial part of the wave function for an ensemble of silver atoms.[90]

The memory **M**, upon interaction with **A**, is supposed to "enter into its book of records" the state of **A** at the end of its measurement

[89] Using the word "memory," I do *not* think of brains. I rather think of the memory of a computer; however, the memory considered in the following is an idealized one, as the records made in it are treated as permanent. (Part of **M** may be a photographic plate, and black spots on it the records.)

[90] There is much reference in the literature to the need for a classical description of the outcome of an experiment, and the necessity of using macroscopic tools for its determination. The macroscopic nature of the instruments is implicitly used by idealizations of the type of Eq. (3): Compare Wigner[7] and our Appendix F, and see our footnotes 38 and 86. Otherwise, the macroscopic nature of **M** or **A** seems to be irrelevant to our reasoning, as we treat here classical states of any gadgets as mixtures of pure quantum states, as outlined in footnote 63. See, however, page 102.

of A upon **O**. All results of any relevant measurements ever made are assumed to be kept in this book in the order (according to **M**'s proper-time parameter) in which they were entered. We now make the important assumption that *these records shall be indelible,*[91] by which we mean that *we shall not guarantee the applicability of quantum theory for any processes which destroy these records.* It is easily seen that this completely solves the measurement problem in quantum mechanics.

In order to see this, we may keep things simple by ignoring all details about anything entered in the past (P) into the memory. We will simply write μ_P for the quantum state of the memories before their interaction with the instruments **A**. Thus, the state of the ensemble $E_{(2)}^{\mathbf{OAM}}$ after a successful[62] interaction of **O** with **A** will be described by us schematically[92] as a pure state

$$\psi_{(2)}^{\mathbf{OAM}} = \mu_P \tilde{\psi}_{(2)}^{\mathbf{OA}} \approx \mu_P \sum c_n \xi_n \tilde{\phi}_n.$$

On account of the interaction between **A** and **M**, this state will now change into[93]

$$\psi_{(3)}^{\mathbf{OAM}} \approx \sum c_n \mu_{P,\mathrm{A}n} \tilde{\xi}_n \tilde{\phi}_n, \tag{15}$$

where $\mu_{P,\mathrm{A}n}$ is a state of **M** in which the record of finding **A** in state ξ_n has been added to the old records P. As $\mu_{P,\mathrm{A}n}$ and $\mu_{P,\mathrm{A}n'}$ for $n \neq n'$ assumedly are eigenfunctions to different[72] eigenvalues of the "memory observables," we now will have

$$\int \mu_{P,\mathrm{A}n}^{*} \mu_{P,\mathrm{A}n'} \, dq_{\mathbf{M}} = \delta_{nn'}. \tag{16}$$

The following considerations will now remain valid even if we would "reunite Wigner's two Stern–Gerlach beams," by which the various $\tilde{\xi}_n$ in (15) would all go over into the same $\xi_{(4)}$. All the same,

[91] Compare P. A. Moldauer, *Bull. Am. Phys. Soc.* **13** (No. 2), 180 (1968), Abstract **ED2**.

[92] In the sense of footnote 63.

[93] Here, $\tilde{\xi}_n$ is related to the ξ_m in a way similar to the relation (4) between $\tilde{\phi}_n$ and the ϕ_m. Note that Eq. (15), like Eq. (3), is an idealization, and correspondingly is all that follows. Compare footnote 90.

when next we measure any observable B upon the system **OA**, the cross terms in

$$\int \psi_{(4)}^{\mathbf{OAM}*} \mathcal{B} \psi_{(4)}^{\mathbf{OAM}} dq_{\mathbf{O}} dq_{\mathbf{A}} dq_{\mathbf{M}}$$

with $n \neq n'$ $\left[\text{after inserting } \psi_{(4)}^{\mathbf{OAM}} = \tilde{\xi}_{(4)} \sum c_n \mu_{P, \mathrm{A}n} \tilde{\phi}_n\right]$ will now vanish on account of (16), so that the density operator describing the state (15) may[94] be replaced for all future purposes by a reducible density operator describing a mixed state with probabilities $|c_N|^2$ for the various reduced pure states

$$\psi_N^{\mathbf{OAM}} = \mu_{P, \mathrm{A}N} \tilde{\xi}_N \tilde{\phi}_N, \tag{17}$$

with the $\tilde{\xi}_N$ replaced by $\tilde{\xi}_{(4)}$ for the ensemble $E_{(4)}^{\mathbf{OAM}}$ if Wigner would reunite the beams after counting them.

The old argument *against* reducibility[95] was that there might exist "measurements of Moldauer's type Z" which would show that the state of $E_{(3)}^{\mathbf{OA}}$ or $E_{(4)}^{\mathbf{OA}}$ would be "merely an *improper* mixture," and that the state of $E_{(3)}^{\mathbf{OAM}}$ could *only* be described by the pure state (15), because supposedly there would exist some observable B such that some matrix elements $\langle \mu_{P, \mathrm{A}n} \tilde{\xi}_n \tilde{\phi}_n | B | \mu_{P, \mathrm{A}n'} \tilde{\xi}_{n'} \tilde{\phi}_{n'} \rangle$ would not vanish for $n \neq n'$. *Here is where the postulate of indelible records enters.* The interaction term in the hamiltonian which would describe the interaction of the measuring apparatus **B** with possibly the whole shebang **OAM** will certainly contain terms depending linearly upon the observable B. In as far as this is an interaction between **B** and **M**, it might enter into **M** a record of the state ζ_m of the apparatus **B**, but it is not to delete an old record (An) of the result of the previous measurement made by **A**. Therefore, the interaction hamiltonian $\mathcal{H}_I$ may have

[94] Note that here we do not claim that the Schrödinger equation would necessarily change the pure-state density operator into the reducible mixed-state operator. Unless such a change is caused by uncertainties in the hamiltonian,[61, 85] we cannot show that we *must* regard $E^{\mathbf{OAM}}$ to be in a "true" mixed state.[75] Here, we *do* show that it is *permitted* to treat the ensemble *as if* it were in a true mixed state. Since reduction of the ensemble simplifies problems of application of the theory, we then *use* this permission.

[95] Compare footnote 75 and the text above it.

matrix elements $\langle \mu_{P,\,An,\,Bm} | H_I | \mu_{P,\,An} \rangle$, but it is not to have matrix elements between $\mu_{P,\,An,\ldots}$ and $\mu_{P,\,An',\ldots}$ for $n \neq n'$. As $\mathcal{B}$ occurs in $\mathcal{H}_I$, $\mathcal{B}$ itself then should not have nonvanishing matrix elements of this kind. This means that we have "postulated away" the possibility of measurements of Moldauer's type Z. This was, of course, the purpose of our postulate in the first place.

We could also have postulated directly that measurements of type Z be not made, granting that the rules about reducibility of a state after a successful measurement will break down when subsequently such measurements of type Z are made, against the warning not to make them. This is the point of view taken by Moldauer.[53] The above discussion simply shows that Everett's 1957 gadget of a memory function may be used handily as a tool for enforcing the ban against measurements of type Z in a plausible way.

In the language of d'Espagnat's 1966 paper,[49] this means that we use the alternative which he mentions on page 829 in his Introduction, of assuming that certain theoretically imaginable observables are in principle not observable.[96] The observables here excluded are those which would change records of previous measurements if they would occur in interaction terms of the hamiltonian. (If they would *not* occur there, the measuring apparatus could not be affected by the value of these "observables.") For all observables different from these excluded ones, the reducible density matrix (8) will predict the same average value as the density matrix (9) obtained by the interaction of **O** with **A**.

If now uncertainties in the hamiltonian cause incoherence between possible results of the measurement,[85] the difference between (9) and (8) is erased altogether, as (9) will be changed into a *distribution* over states (9) *averaging* to give (8), while (8) itself will remain what it was. What the preceding paragraphs did achieve was ensuring reducibility even before this incoherence sets in.

[96] Again, we may water down this claim to Moldauer's statement that the predictions of conventional objective quantum theory obtained by rule **F5** are consistent with the other parts of this theory, provided that we do not apply the theory to those theoretically imaginable observables which we here are excluding.

1.22. Probability distribution for follow-up measurement

In the preceding section we justified the use of a reducible state for describing the ensemble of object and apparatus at the end of a successful measurement completed by indelible registration of whatever its result may have been. We justified it by showing that it yields the correct mean value of whatever observable B is measured next. We shall now show that it also gives the correct probabilitity distribution over the various possible outcomes of this follow-up measurement.

Let the "actual state" of $E_{(3)}^{\mathbf{OAM}}$ be given by (15), and let χ_m be the orthonormalized eigenfunctions of the observable B. We will consider two separate cases: (i) B is measured upon $\mathbf{O}$ alone; and, (ii) B is measured upon the composite system $\mathbf{OA}$.

Case (i)

In $q_{\mathbf{O}}$-space we expand

$$\tilde{\phi}_n = \sum_m d_{nm}\,\chi_m \qquad \text{with} \qquad \sum_m |d_{nm}|^2 = 1, \tag{18}$$

and in $q_{\mathbf{A}}$-space we expand

$$\tilde{\xi}_n = \sum_l h_{nl}\,\xi_l \qquad \text{with} \qquad \sum_l |h_{nl}|^2 = 1. \tag{19}$$

According to the customary generalization of the rule **F5** to probabilities for a partial system in terms of a state vector for an ensemble of composite systems, the probability of finding $\mathbf{O}$ in the eigenstate χ_m of B after the measurement of A then is

$$\begin{aligned} w_{(3)m}^{\mathbf{B}} &= \iint dq_{\mathbf{A}}\, dq_{\mathbf{M}} \left| \int dq_{\mathbf{O}}\, \chi_m^*\, \psi_{(3)}^{\mathbf{OAM}} \right|^2 \\ &= \iint dq_{\mathbf{A}}\, dq_{\mathbf{M}} \left| \sum_n c_n\, \mu_{P,\,\mathrm{A}n}\, \tilde{\xi}_n\, d_{nm} \right|^2 \\ &= \sum_n \sum_{n'} c_n^*\, c_{n'}\, d_{nm}^*\, d_{n'm} \langle \mu_{P,\,\mathrm{A}n} | \mu_{P,\,\mathrm{A}n'} \rangle \langle \tilde{\xi}_n | \tilde{\xi}_{n'} \rangle. \end{aligned} \tag{20}$$

Now use first[72]

$$\langle \mu_{P,\mathbf{A}n} | \mu_{P,\mathbf{A}n'} \rangle = \delta_{nn'}, \tag{21}$$

and then use

$$\langle \tilde{\xi}_N | \tilde{\xi}_N \rangle = \sum_l \sum_{l'} h^*_{Nl} h_{Nl'} \langle \xi_l | \xi_{l'} \rangle = \sum_l |h_{Nl}|^2 = 1. \tag{22}$$

This gives

$$w^{\mathrm{B}}_{(3)m} = \sum_n |c_n|^2 |d_{nm}|^2. \tag{23}$$

Case (ii)

This time, we expand in $q_{\mathbf{O}}, q_{\mathbf{A}}$-space

$$\tilde{\xi}_n \tilde{\phi}_n = \sum_m d_{nm} \chi^{\mathbf{OA}}_m. \tag{24}$$

We obtain

$$\begin{aligned} w^{\mathrm{B}}_{(3)m} &= \int dq_{\mathbf{M}} \left| \int\int dq_{\mathbf{O}}\, dq_{\mathbf{A}}\, \chi^{\mathbf{OA}*}_m \psi^{\mathbf{OAM}}_{(3)} \right|^2 \\ &= \int dq_{\mathbf{M}} \left| \sum_n c_n \mu_{P,\mathbf{A}n} d_{nm} \right|^2 \\ &= \sum_n \sum_{n'} c^*_n c_{n'} d^*_{nm} d_{n'm} \langle \mu_{P,\mathbf{A}n} | \mu_{P,\mathbf{A}n'} \rangle \\ &= \sum_n |c_n|^2 |d_{nm}|^2, \end{aligned} \tag{25}$$

which is the same as (23) in Case (i).

Both results agree with the prediction obtained by replacing the state $\psi^{\mathbf{OAM}}_{(3)}$ of $E^{\mathbf{OAM}}_{(3)}$ by the reducible (mixed) state with probability $|c_N|^2$ for an ensemble $E^{\mathbf{OAM}}_N$ or $E^{\mathbf{OA}}_N$ in a pure state $\psi^{\mathbf{OAM}}_N$ [Eq. (17)] or $\psi^{\mathbf{OA}}_N = \tilde{\xi}_N \tilde{\phi}_N$. If $\{\phi^{\mathbf{N}}_n\}$ is a complete set of states of the part **N** of **OAM** not affected by the measurement of B, the same results for $w^{\mathrm{B}}_{(3)m}$ can also be obtained by $w^{\mathrm{B}}_{(3)m} = \sum_n \left| \int\int\int dq_{\mathbf{O}}\, dq_{\mathbf{A}}\, dq_{\mathbf{M}}\, \phi^{\mathbf{N}*}_n \chi^{\mathbf{Q}*}_m \psi^{\mathbf{OAM}}_{(3)} \right|^2$, where $\mathbf{QN} \equiv \mathbf{OAM}$.

1.23. Visual recording

Sometimes, the interaction between **A** and **M** is not direct. The apparatus **A** may interact with the photon field **L**, initially in a state λ_0. The interaction may change the state of the ensemble $E^{\mathbf{AL}}$ according to[97]

$$\xi_n \lambda_0 \to \tilde{\xi}_n \lambda_n \qquad \text{with} \qquad \tilde{\xi}_n = \sum_l h_{nl}\, \xi_l, \qquad \sum_l |h_{nl}|^2 = 1. \quad (26)$$

Next, the modified field **L** may interact with the memory **M** by[97]

$$\lambda_n \mu_P \to \tilde{\lambda}_n \mu_{P,\mathrm{A}n} \qquad \text{with} \qquad \tilde{\lambda}_n = \sum_k C_{nk}\, \lambda_k, \quad \sum_k |C_{nk}|^2 = 1. \quad (27)$$

We assumed here that through the **L** field we still can record in **M** *which* apparatus **A** caused the light signal λ_n to travel to **M**.

We thus obtain from $\psi_{(2)}^{\mathbf{OALM}} = \lambda_0\, \mu_P\, \psi_{(2)}^{\mathbf{OA}}$ a state

$$\psi_{(4)}^{\mathbf{OALM}} = \sum_n c_n\, \mu_{P,\mathrm{A}n}\, \tilde{\lambda}_n\, \tilde{\xi}_n\, \tilde{\phi}_n. \quad (28)$$

This differs from $\psi_{(3)}^{\mathbf{OAM}}$ [Eq. (15)] merely by containing $\tilde{\lambda}_n \tilde{\xi}_n$ instead of $\tilde{\xi}_n$, or, by containing $h_{nl} \sum_k C_{nk} \lambda_k$ instead of h_{nl}. Then, where we used $\sum_l |h_{nl}|^2 = 1$ in (22), we now would use

$$\sum_l h^*_{Nl} \sum_k C^*_{Nk}\, h_{Nl} \sum_{k'} C_{Nk'} \langle \lambda_k | \lambda_{k'} \rangle = \sum_l |h_{Nl}|^2 \sum_k |C_{Nk}|^2 = 1.$$

Our conclusions remain the same.

1.24. Everett's "relative states"

We have quoted Everett's paper[10] in the above, because his memory wave function[98] was useful to us for schematizing how different results of successful measurements correspond to states that are

[97] This is an idealization, like Eq. (3) was. Compare footnotes 63, 90, and 93.
[98] For our $\mu_{\ldots}$, Everett[10] wrote $\psi^0_{[\ldots]}$.

mutually orthogonal and between which future measurements have no transition matrix elements.

Another formal idea discussed by Everett is the concept of what he calls[99] "relative" quantum states.

Consider an ensemble of composite systems **AB**. We may expand its state $\psi^{\mathbf{AB}}$ in terms of a complete orthonormal set $\{\psi_m^{\mathbf{A}}\}$ of states of **A** alone, by

$$\psi^{\mathbf{AB}} = \sum_m c_m \, \psi_{(\mathbf{A})m}^{\mathbf{B}} \, \psi_m^{\mathbf{A}}, \tag{29}$$

where the $\psi_{(\mathbf{A})m}^{\mathbf{B}}$ are normalized (but generally not orthogonal) vectors in **B**'s Hilbert space.[101]

In the state $\psi^{\mathbf{AB}}$ of $E^{\mathbf{AB}}$, Everett then calls $\psi_{(\mathbf{A})m}^{\mathbf{B}}$ the *relative state* of **B** *corresponding to* the state $\psi_m^{\mathbf{A}}$ of **A**:

$$\psi_{(\mathbf{A})m}^{\mathbf{B}} = \psi(\mathbf{B}; \text{rel.}\, \psi_m^{\mathbf{A}}, \mathbf{A}). \tag{30}$$

Thus, $\psi_{(\mathbf{A})m}^{\mathbf{B}}$ is obtained "in **B**-space" by normalization of the coefficient $\int dq_{\mathbf{A}} \, \psi_m^{\mathbf{A}*} \, \psi^{\mathbf{AB}}$ in the expansion of $\psi^{\mathbf{AB}}$ in terms of *any* orthonormal set "in **A**-space" of which $\psi_m^{\mathbf{A}}$ is a member.

If, for a given state $\psi^{\mathbf{OA}}$ of an ensemble $E^{\mathbf{OA}}$ of composite systems **OA**, we choose for the state $\psi_m^{\mathbf{A}}$ an observed "pointer position" state ξ_m of the "apparatus" subsystem **A** used for completing a measurement upon the other ("object") subsystem **O**, then $\psi_{(\mathbf{A})m}^{\mathbf{O}} = \psi(\mathbf{O}; \text{rel.}\, \xi_m, \mathbf{A})$ is exactly what we have called the *reduced state* $\tilde{\phi}_m$ of **O** after

[99] Correlations between $\psi_m^{\mathbf{A}}$ of **A** and $\psi_{(\mathbf{A})m}^{\mathbf{B}}$ of **B** as in the state (29) of $E^{\mathbf{AB}}$ are, of course, not new. Without giving it a specific name, such correlations have been studied ever since the Einstein–Podolsky–Rosen paradox[100] was published, and probably before that time.

[100] A. Einstein, B. Podolsky, and N. Rosen, *Phys. Rev.* **47,** 777 (1935). See also our Preface.

[101] If $\{\psi_n^{\mathbf{B}}\}$ is a complete orthonormal set of states of **B**, we may expand $\psi^{\mathbf{AB}} = \sum_m \sum_n c_{mn} \, \psi_m^{\mathbf{A}} \, \psi_n^{\mathbf{B}}$. Then, $\psi_{(\mathbf{A})M}^{\mathbf{B}} = |c_M|^{-1} \sum_n c_{Mn} \, \psi_n^{\mathbf{B}}$, with $|c_M|$ given by $|c_M|^2 = \sum_n |c_{Mn}|^2$.

the measurement. Or, after the result of the measurement has been recorded in **M**, we find the reduced state $\psi_N^{\mathbf{OA}}$ of the ensemble $E_N^{\mathbf{OA}}$ of **OA** systems, by finding the coefficient of the state $\mu_{P,\mathrm{A}n}$ in the expansion (15) of the state vector $\psi_{(3)}^{\mathbf{OAM}}$ of $E_{(3)}^{\mathbf{OAM}}$, and then normalizing it by omitting the factor c_N:

$$\psi_N^{\mathbf{OA}} = \psi_{(\mathbf{M})N}^{\mathbf{OA}} = \psi(\mathbf{OA}; \text{rel. } \mu_{P,\mathrm{A}N}, \mathbf{M}) = \tilde{\xi}_N \tilde{\phi}_N. \qquad (31)$$

Thus, the "relative state" formalism may be a handy tool for writing down reduced states *when it works*. There is, however, no *a priori* guarantee that a relative state could always be a possible reduced state, because the formalism contains no verification whether or not (for calculating predictions for future measurements) it is allowable to treat the state ψ of the original composite system as a mixed state.[102] In order to provide this guarantee, one could confine use of this formalism to states relative to memory states, as in Eq. (31), for state vectors of the type of $\psi_{(3)}^{\mathbf{OAM}}$.

The beauty of describing the "result" of the measurement of A upon an ensemble by an unreduced state vector like $\psi_{(3)}^{\mathbf{OAM}}$ [of Eq. (15)] lies in that it is completely determined by the measurement, and independent of our inability to predict the specific result A_N of a measurement upon an individual system **O** in the ensemble. Nevertheless, (31) shows that $\psi_{(3)}^{\mathbf{OAM}}$ contains all that is necessary for making predictions for any subensemble corresponding to any selected or observed record A_N in the memory. *In Everett's words*,[103] state vectors of the type of $\psi_{(3)}^{\mathbf{OAM}}$ give 'a representation in terms of a *superposition*, each element of which contains a definite observer state and a corresponding system

[102] For instance, if we do *not* record the result of the A measurement before making upon **OA** a B measurement of Moldauer's type Z,[75] we know that it is wrong to replace $\psi_{(2)}^{\mathbf{OA}}$ by a probability distribution over reduced states. (See Secs. 1.18 and 1.20.) Yet, $\psi_N^{\mathbf{O}} = \tilde{\phi}_N$ would be the relative state $\psi(\mathbf{O}; \text{rel. } \xi_N, \mathbf{A})$.

[103] Reference 10, page 459. *Here, we merely quote*. For comments, see Secs. 1.25 through 1.27.

state.[104] Thus with each succeeding observation (or interaction), the observer state "branches" into a number of different states.[104] Each branch represents a different outcome of the measurement and the *corresponding* eigenstate for the object-system state.[104] All branches exist simultaneously in the superposition after any given sequence of observations.'[103]

1.25. Is there a state vector for the universe?

In Sec. 1.5, we preceded our discussion of ensembles of systems **O**, **A**, etc., by the remark that such discussions tacitly assume that the state of ensembles of these systems can be discussed separate from the state of the remainder of the universe.

If, however, two systems interact, like **O** and **A** in Sec. 1.17, a state vector $\psi^{\mathbf{OA}}$ which originally might have been factorizable into state vectors for **O** and for **A** separately as in Eq. (2), becomes a superposition as in Eq. (3) that is no longer factorizable. Such a state describes correlations between **O** and **A**, and these correlations are not fully describable classically when there exist observations B which show interference effects. In that case we were not allowed to regard the ensemble $E^{\mathbf{OA}}$ as being in a true mixed state, and the statement that there would be a probability $|c_n|^2$ for **O** to be in state $\tilde{\phi}_n$ while **A** would be in state ξ_n is not a complete description of the situation. In other words, it is not really allowable in this case to talk about the state of **O** alone or of **A** alone, or even about probabilities for such states. We must regard **O** and **A** as *inseparable*.

However, at *some* time in the past, *any* part of the universe has interacted with some other part, and therefore it would seem as if we would be forced to reject any talk about a state of some part of the universe, and that only the universe as a whole would be in one state or another. This was, essentially, Everett's conclusion.[10]

[104] In $\psi_{(3)}^{\mathbf{OAM}}$, the "observer states" are the $\mu_{P,\,An}$ with "corresponding object-system states" $\tilde{\xi}_n\tilde{\phi}_n$ obtainable from ψ as relative states as in Eq. (31). After a next observation of B, the observer states will be $\mu_{P,\,An,\,Bm}$, and so on. The "branching" refers to the increasing number of indices on the observer states $\mu_{\dots}$. (This footnote is not a part of the quotation.)

Since, however, state vectors describe probabilities in ensembles, our next conclusion is that the state vector of the universe is a meaningless quantity unless there exists an *ensemble of universes* on which quantum-mechanically predicted probability distributions can be verified experimentally.

However this may be, in this type of a theory one avoids reduction of state vectors [like the reduction from Eq. (15) for $\psi_{(3)}^{\mathbf{OAM}}$ to some particular $\psi_N^{\mathbf{O}}$ for the partial system **O**], as it is here assumed to be meaningless to talk about state vectors for ensembles of partial systems.

With a state vector ψ describing an ensemble of entire universes rather than of single systems **O**, one needs, of course, a formula somewhat different from the old interpretative rule **F5** of Sec. 1.11, for obtaining from ψ the probabilities for definite outcomes of definite measurements on an object **O** in this universe. Everett therefore replaces the interpretative rule **F5** for $\psi_0^{\mathbf{O}}$ of Eq. (1), by an interpretative rule for the final universal state vector $\psi^{\mathbf{OAB}\ldots\mathbf{M}\ldots}$ after several measurements have been made. This final state is expanded in terms of products, of the various possible states $\mu_{P,\,An,\,Bm,\ldots}$ of the memory **M** used for recording the results of the measurements upon **O**, each with its corresponding relative state $\psi(\mathbf{OAM}\ldots;\text{rel. }\mu_{P,\,An,\,Bm,\ldots},\mathbf{M})$. The absolute squares of the coefficients $c_{P,\,An,\,Bm,\ldots}$ in this expansion, properly summed over all possibilities for the irrelevant past P, then are postulated to be the probabilities (in the ensemble of universes) that $A = A_n$, $B = B_m$, ... were the results of the measurements made. Everett[10] in his chap. 5 explains why any probability postulate different from this one would violate conservation of the sum of probabilities.

It still would be possible to consider subensembles of our ensemble of universes, in as far as we could reduce $\psi_{(3)}^{\mathbf{OAM}}$ to $\psi_N^{\mathbf{OAM}}$ by selecting the subensemble in which a particular pointer position has been indelibly recorded in the memory **M**. We have seen, however, that, even with the conventional probability interpretation of state vectors, this reduction is *not really necessary for calculating conditional probabilities* like that for $B = B_m$ following $A = A_n$ as we did in footnote

74 and in Sec. 1.22, nor did we need it for calculating the ensemble average of B after measuring A as in Sec. 1.21. Also, the reduction $E_{(3)}^{\mathbf{OAM}} \to E_N^{\mathbf{OAM}}$ is performed in the conventional theory mainly as a preliminary to a further reduction to $E_N^{\mathbf{O}}$, and the latter step was rejected by Everett. Therefore, he considered it simpler to reject the reduction formalism altogether.

It seems rather strange to talk about an ensemble of universes, as there is only one single universe we are acquainted with. This problem was solved by Everett in a simple way, by remarking that the different members of an ensemble never interact with each other, so that obviously we cannot be aware of the other members of the ensemble.[105, 106]

While reading Everett's paper in the original, you will notice that his language somewhat deviates from the wording which we here have used in an attempt to stay within objective quantum theory.[107] Everett talks about the "state vector of the universe", omitting the word *ensemble*. This makes his language reminiscent of Schrödinger's cat, and makes his talk about the "branching of the universe" reminiscent of the "suspended animation" of Wigner's friend. In order to avoid such misunderstandings, *we introduced the word "ensemble" e x p l i c i t l y in our presentation of Everett's theory.*

1.26. Objections against Everett's model

In Everett's theory, as interpreted by us in the language of objective quantum theory, the essential point is that the universe as we know it is just one arbitrary member of an entire ensemble of

[105] See footnote on pages 459–460 of reference 10. Everett there stresses the *reality* of the unobservable "branches of the universe" (members of the ensemble of universes). The meaning of the word "reality" in this context is not clear.

[106] B. S. DeWitt, *The Everett–Wheeler Interpretation of Quantum Mechanics*, chap. xii of *Battelle Rencontres (1967)* edited by C. M. DeWitt and J. A. Wheeler. (W. A. Benjamin, New York, 1968.) See in particular pages 325–326.

[107] Everett also confined his considerations to idealized measurements only; that is, the special case characterized by our Eq. (5). However, it is easy to generalize his discussions.

universes, and that the laws of physics are probability laws for this big ensemble, and therefore are not deterministic for "our own" universe by itself.

It is impossible to say that this point of view would be definitely wrong (*i.e.* illogical). Most people "do not like it." We will here summarize the main objections made.

The simplest and possibly the most important objection is that quantum theory then would describe something unobservable, and therefore something useless to talk about. As we cannot observe those other members of the ensemble of universes, if they exist at all, why bother to talk about them? As Shimony[44] well expressed this in a footnote to his paper, this would be the worst violation imaginable of the principle of Occam's Razor: *Essentia non sunt multiplicanda praeter necessitatem* (which could be translated rather freely into "Keep things simple!").

In the second place, if it were true that the universe would "branch" at any time when a new record of an as yet undisclosed measurement result is entered into the World Memory Bank **M**, this Bank would seem to rule the world's "growth" in a hard-to-justify manner.

In the third place, the main *purpose* of envisaging the "existence" of the other branches of the universe seems to be an attempt at escaping from a world picture in which the outcomes $A = A_N$ of measurements are chosen by nature in an unexplainable way, and it seems to me that such attempts are futile.[108]

[108] In reply to the claim of conventional quantum theory that there is no law of nature for choosing the value of N in the result $A = A_N$ of a measurement of A, Einstein has said that he did not like the idea that God would be playing dice.[109] Though it may seem that leaving all "branches" as terms in the universal state vector ψ would *eliminate* the making of this choice of N, in fact for any living person it merely *postpones* this choice until the time when one uses ψ *relative to* a state of **M** labeled by the N corresponding to the measurement result of which one "is aware." This N, however, was not predictable by this person before he "observed" its unpredictable value. As a physicist, what would he care about hypothetical other branches of the universe? As they are not observable, he cannot verify anyhow Everett's claim that N in them may have different values! If $N = 17$ in *our* branch, then, what kind of lottery made N not 17 in somebody else's branch, and 33 in ours? In assigning the "me" whom I know, to one of the branches of the universe, why

In the fourth place, Everett's claim that his ψ would describe the entire world is a hoax, and *fortunately* it is a hoax. Describing the entire universe by a single state vector is, of course, practically impossible. What Everett does is writing "dot-dot-dot" for whatever is irrelevant to some particular discussion. This is equivalent to claiming that, after all, it is possible to describe what is going on in the laboratory without being explicit about the state of the last electron in the Andromeda Nebula (included in the dot-dot-dot). Then, one might as well cut the state vector to the bone, and reduce it to things like our $\psi^{\mathbf{OABM}}$, or even our initial $\psi_0^{\mathbf{O}}$ for the (ensemble of systems) **O** alone. But, if we start at time t_0 with some $\psi_0^{\mathbf{O}}$, this $\psi_0^{\mathbf{O}}$ must have been obtained from some earlier $\psi^{\mathbf{OZ}}$ by reduction, if **Z** was the apparatus of the preceding interaction with **O**. Thus, as long as *this* reduction (by "dot-dot-dot") *was* permitted, we might as well use reductions *whenever* we can.

Finally, it simply is not true that the objects studied in physics would be inseparable because of their previous interactions. We have stated before that any physical theory has limits of validity.[29] It is wise to keep these limits always in mind. For one thing, a physicist is not going to make measurements for the purpose of destroying the last record of a previous result. Therefore, the limitation which we have set ourselves in Sec. 1.21 by postulating the indelibility of the records is an acceptable one. With this restriction to the assumed applicability of the theory, however, we have seen in Secs. 1.21–1.22 that the reasons for unseparability of the objects of the physicists' interest (like **O** and **A** and **M**) disappear where we are ready for a reduction, even without having to wait for the soon-entering incoherence between observed states.

Therefore, while it may be impossible to prove that Everett's interpretation of quantum theory would be a logical impossibility, there are many reasons for nevertheless rejecting it. But this does not mean

---

did God assign me to the branch with $N = 17$ and not 33? Was God playing dice about *me*? If God chooses *our* value of N, can He do so only by playing dice?

[109] See, for instance, remarks by M. Born and by N. Bohr on pages 176 and 218 of reference 77.

that we could ignore Everett's paper. Beside making fascinating reading, it did suggest already in 1957 a way for avoiding objections against the conventional "mixed-state" interpretation of the results of successful measurements, by the tool of the indelible memory state. Moreover, it contains an interesting analysis of the *reasons* for the customary probability postulates of quantum theory, which analysis, too, is based upon the assumption of indelibility of the records entered into the memory **M**.[110]

1.27. Conclusion

We conclude that, after all, reductions of states after successfully recorded measurements are permisssible. While in Sec. 1.5 we simply postulated that it made sense to talk about quantum-mechanical states of ensembles of limited physical systems, and while in Sec. 1.11 we introduced the statistical postulate **F5** without verifying its consistency and compatibility, we have seen in Secs. 1.21–1.22 why these assumptions usually are justified. (We do not try to generalize these assumptions to cases where they obviously are not justifiable.)

This gives, in the first place, a justification for the conventional formalism for the "preparation" of the initial state of an ensemble of systems. If we want to prepare the eigenstate ϕ_N corresponding to $A = A_N$, we first make an ideal measurement of A. The result is describable by Eq. (15), with $\tilde{\phi}_n = \phi_n$ according to Eqs. (4)–(5). That is, we obtain

$$\psi_{(3)}^{\mathbf{OAM}} = \sum c_n \, \mu_{P,\mathrm{A}n} \, \tilde{\xi}_n \, \phi_n . \tag{32}$$

But, as we proceed to describe the measurement of some other observable B upon **O** prepared by the first measurements, there is no purpose in cluttering up further discussions and calculations by terms with $n \neq N$ while the preparation consists of rejecting from the ensemble of further measurements all cases with $\mu_{P,\mathrm{A}n} \neq \mu_{P,\mathrm{A}N}$. There also is no purpose in cluttering later calculations by past memory records P, or by perturbed pointer position states $\tilde{\xi}_n$ of

[110] Reference 10, sec. 5.

previously used apparatus **A**. In short, we do not want to carry along all garbage contained in (32) that is irrelevant to later measurements; so, we first pick from $E_{(3)}^{\mathbf{OAM}}$ the subensemble $E_N^{\mathbf{OAM}}$ not only experimentally, but also in the calculations, and then we simplify $\psi_N^{\mathbf{OAM}}$ to $\psi_N^{\mathbf{O}} = \phi_N$, as descriptive of the ensemble $E_N^{\mathbf{O}}$ of systems **O** lifted from the systems **OAM** in $E_N^{\mathbf{OAM}}$.

Thus, it is natural to *start* every calculation with a *reduced* state vector. Then, it is also only natural to express the *final* state in a similar way, ready to be used as initial state for whatever measurement we want to make next. That is, if we "throw out the garbage" in the *initial* state, there is no reason for saving all garbage that accumulates later while we proceed to the *final* state.

The garbage is removed by restricting our discussions to the best-defined subensemble that is consistent with the problem at hand and that is descriptive of the conditions initially imposed experimentally, which may include selecting definite results of preceding measurements. Obviously, incomplete preceding measurements do not impose such conditions. This eliminates the danger of suppressing in the theoretical description (by unallowable reductions) interference effects that the experimental setup allows.

Where we can (and therefore probably do) perform a reduction of an ensemble (and, thus, of its state), we feel no urge to use fancy language by saying that "the universe has branched" and that we are riding one of its branches (like one may be riding a train). However, when we meet other people who like to use this fancy language, we might keep in mind that in practice the results of their calculations are exactly the same as the ones we obtain. They just make the derivations sound more impressive, and probably are carrying along some of the garbage that we threw out, just for justifying their language, but not affecting the results of their work.

Leaving them alone in their endeavor, we recognize how in Secs. 1.21–1.22 our task was lightened by borrowing from them the formalism of memory state vectors, as a tool for schematizing the reasons for the consistency of the conventional statistical interpretation of quantum theory as advocated by the Copenhagen School.[14]

Perhaps we admire the placidity with which this School has always accepted the probability rule **F5** as an axiom. We feel, however, happy about knowing something about the area of applicability and the limits of validity of this rule, suspecting all along that the people in Copenhagen have been well aware of these limits.

If then somebody constructs some special case in which he steps outside this area of applicability, transgressing these limits of validity, we do not become alarmed and do not question all of quantum theory. We rather try to find out how to replace routine treatment in such a case by an adequate alternative consistent with the fundamentals of the conventional theory. Thus we agree that in Wigner's 1963 example (Sec. 1.20) the state vector should not be reduced, but feel no need for revolutionary action on account of it.

CHAPTER 2

RETRODICTION IN OBJECTIVE QUANTUM THEORY

2.1. Introduction

In 1964 an interesting paper on time reversal in quantum theory was published by Aharonov, Bergmann, and Lebowitz,[111] in the following quoted as ABL. In this Chapter 2 we shall give a schematized account of their results. We shall somewhat generalize their discussion of sequences of idealized measurements, by considering also nonideal measurements, and interactions with other instruments or gadgetry between consecutive measurements. This will somewhat help the understanding of reasons for the lack of time symmetry in the applicability of the probability rules of quantum theory. We shall also show how wrong applications of these rules have led some authors to unacceptable conclusions.

We understand quantum theory here as the objective theory discussed in Chapter 1. There we have seen that the reduction of state vectors in conventional quantum theory is justifiable. Here we shall take this result for granted, and we shall no longer go through the formalism of the theory of measurements for demonstrating it. That is, the state vector of the apparatus will not appear on the following pages. Instead, we consider the state vectors before a measurement and after the reduction following that measurement, as state vectors for two different ensembles.

For the sake of simplicity, we treat in the following only the special case that all observables measured have only a finite number of eigenvalues, all discrete and nondegenerate.

2.2. Use of the conventional postulates of quantum theory

I shall follow Aharonov, Bergmann, and Lebowitz,[111] starting by temporarily taking for granted the conventional interpretation of quantum theory, which includes the following ideas:[112]

(1) State vectors and density operators describe *ensembles*,[113] and allow us to make probability predictions about them.

(2) After successful measurements of an observable A have been made upon all members of an ensemble E of systems **O** (the *objects* of the measurements), it is possible to describe the final state of this ensemble as a mixed state. If E originally was in a pure state $\psi = \sum c_i \phi_i$, where $\mathcal{A}\phi_i = A_i \phi_i$, the ensemble E in its final state, according to the *reduction principle*, can be subdivided into subensembles E_i. The numbers N_i of member systems **O** in these subensembles are fractions $|c_i|^2$ of the number N of systems **O** in E.

The compatibility of the reduction principle with the remainder of quantum theory has seriously been questioned by a large number of authors.[114] We discussed this problem in Chapter 1. Here, we simply use the result of that discussion,[115] that notwithstanding all of this criticism there still is a justifiable place for reduction of state vectors in quantum theory.

If the measurement is an *ideal* one, each ensemble E_i according to von Neumann's projection postulate is in a pure state ϕ_i. This requires

[111] Y. Aharonov, P. G. Bergmann, and J. L. Lebowitz, *Phys. Rev.* **134,** B 1410 (1964), in the following quoted as ABL.

[112] For a critical discussion of these ideas, see Chapter 1.

[113] See Secs. 1.3, 1.8, and 1.12.

[114] For a bibliography, see B. S. De Witt and R. N. Graham, *Am. J. Phys.* **39,** 724 (1971).

[115] Validity of some of the concepts used in this discussion required that the measuring apparatus be very large compared to the objects on which the measurements are made. See footnotes 66 and 68 on page 28.

that the measurement shall not disturb the observable measured, so that it shall be reproducible. (That is, repetition of the measurement upon E_i in absence of intervening interactions shall reproduce the same old result A_i for all members of E_i, assuming, of course, that A is a conserved quantity. Compare page xvi.) In other words: An ideal measurement *prepares a state*.

In order to cope with the case that one would measure observables that are *not* conserved, one may construct different observables, that *are* conserved and that coincide with the quantities measured at the time of the measurement. (For instance, on a freely moving particle one might measure an observable like its position $\mathbf{x}(t)-\mathbf{v}t$ at time zero, for which the Schrödinger operator would be $\mathbf{x}-t\mathbf{p}/m$.) In the following we shall always tacitly assume that all quantities measured are conserved quantities, so that we avoid problems about change of observables while they are being measured.

ABL[111] discuss also *sequences* of different ideal measurements, in which the initial state for a subsequent measurement is assumed to be identical with the final state prepared by the preceding measurement.

It is desirable to generalize their considerations by considering also nonideal measurements, and by considering also the possibility that a following measurement is preceded by interaction of the objects **O** with other objects, including those used to set the stage for the following measurement. As these other objects usually will be in mixed states, it should be expected that these interactions will transform pure states of the **O** into mixed states (after separation of the **O** from the other objects by another reduction of state). The same holds for the states of the subensembles E_i emerging from a measurement in which the apparatus (in a mixed state itself) perturbs the states of the quantity measured.[116]

A complication often met is that the objects **O** in measurements and in interactions with other objects will lose their original character. The **O** may be particles which are absorbed during the measurement,

[116] See the example in Appendix D.

and what is left may be the absorbing particle in an excited state. Or, the measurement or later interaction may include a nuclear, atomic, or chemical reaction. Therefore, we should allow the final ensemble to consist of particles different in nature from the original particles **O** on which the measurement was started, and the final state will describe this modified ensemble.

In Sec. 2.9 we will discuss what additional assumptions are necessary, if the predictive probability interpretation of state vectors described under (2) above is replaced by the less powerful but more time-symmetric postulate Eq. (49) of Sec. 2.7.

2.3. Resolution of mixed states

We may resolve any mixed state into a set of component pure states occurring with certain probabilities. This decomposition of a mixed state in quantum theory is not unique. There are two different ways in which we can approach the task of taking away the ambiguity of such a decomposition. The one takes a mathematical approach, and the other takes a physical approach. The two methods lead to entirely different results. This is best illustrated by some examples.

Example 1.

Let ϕ_{n+} and ϕ_{n-} be the two spin states for a silver atom corresponding to $\mathcal{S}_n \phi_{n\pm} = (\pm\hbar/2)\,\phi_{n\pm}$, where $\mathcal{S}_n = \mathbf{n}\cdot\mathbf{S}$ = the spin in the direction **n**. Take **n** at an angle θ with the z-axis and $(90° - \theta)$ with the x-axis. Then,

$$\phi_{n+} = c\,\phi_{z+} + s\,\phi_{z-} \quad \text{and} \quad \phi_{n-} = -s\,\phi_{z+} + c\,\phi_{z-}$$

where $c = \cos(\theta/2)$ and $s = \sin(\theta/2)$.

Let $W_{n\pm}$ be the density operators for the pure states $\phi_{n\pm}$. If $\gamma = \cos\theta$ and $\sigma = \sin\theta$, we find, in the representation which diagonalizes $W_{z\pm}$,

$$2\,W_{z+} = \begin{pmatrix} 2 & 0 \\ 0 & 0 \end{pmatrix}, \qquad 2\,W_{x+} = \begin{pmatrix} 1 & 1 \\ 1 & 1 \end{pmatrix},$$

$$2\,W_{n+} = \begin{pmatrix} (1+\gamma) & \sigma \\ \sigma & (1-\gamma) \end{pmatrix}, \qquad 2\,W_{n-} = \begin{pmatrix} (1-\gamma) & (-\sigma) \\ (-\sigma) & (1+\gamma) \end{pmatrix}.$$

Now assume that by a Stern–Gerlach apparatus we measure S_z on a beam of silver atoms traveling in the y-direction. In the $S_z = -\hbar/2$ beam we place a gadget that turns the spin of every atom in it from $S_z = -\hbar/2$ to $S_x = +\hbar/2$. Then, we mix this beam incoherently with the unaltered $S_z = +\hbar/2$ beam.

Before entering the Stern–Gerlach apparatus, the beam may have been polarized by taking for it the $S_{n'} = +\hbar/2$ beam from a preceding Stern–Gerlach apparatus, so that the initial probabilities for $S_z = \pm\hbar/2$ were given by $p_{z+} = \cos^2(\theta'/2)$ and $p_{z-} = \sin^2(\theta'/2)$, functions of the angle θ' of $\mathbf{n}'$ with the z-axis.

The density matrix of the final beam is now $W' = p_{z+} W_{z+} + p_{z-} W_{x+}$.

The *physical* resolution of this mixed state is, as given by the latter formula, into pure states ϕ_{z+} and ϕ_{x+} determined by the gadgetry used in the experiment, and independent of p_{z+} and p_{z-}, that is, independent of the polarization θ' of the incident beam. We notice, however, that ϕ_{z+} and ϕ_{x+} are not orthogonal, and that it is a mere happenstance that they form a complete set of spin states. (See Examples 2 and 3 for cases where the physically preferred pure components of the mixture form incomplete or overcomplete sets.)

Mathematically, we can resolve W' into a mixture of pure states which are eigenstates of one single observable S_n, and which therefore *form a complete orthonormal set*. For this, we choose the angle θ of $\mathbf{n}$ by $\tan\theta = \tan^2(\frac{1}{2}\theta')$ in such a way that

$$[\gamma/(\gamma+\sigma)]\, W_{z+} + [\sigma/(\gamma+\sigma)]\, W_{x+} = W' = \tilde{p}_{n+} W_{n+} + \tilde{p}_{n-} W_{n-},$$

where $\tilde{p}_{n\pm} = \frac{1}{2}[1 \pm (\gamma+\sigma)^{-1}]$.

In general, *any* density matrix W may be resolved in such a way, since W is hermitian and therefore may be diagonalized by a unitary transformation from whatever was the originally chosen basis of the matrix representation of W (the $\phi_{z\pm}$ in the above example) to a properly chosen new basis (here the $\phi_{n\pm}$); the diagonal elements after the diagonalization then are the probabilities in the mixture, and the new basis functions form the complete orthonormal set of pure states.

We see, however, that the $\phi_{n\pm}$ through θ depend upon the $p_{z\pm}$. That is, *this mathematically nice set of pure states is not determined by the apparatus of the measurement and spin turn alone, and therefore cannot be considered representative of the* p r o c e d u r e *by which we mixed the states. Instead,* $\phi_{n\pm}$ *depends additionally upon the polarization of the beam* i n c i d e n t *upon the instruments.*

Example 2.

We could change the apparatus of Example 1 by rotating the $S_z = -\hbar/2$ atoms to $S_z = +\hbar/2$ instead of to $S_x = +\hbar/2$. Then, as far as the spin is concerned, the resulting beam would be in a pure state with $W' = W_{z+}$. We find here an example of a state which was mixed before the spin flip $[W = p_{z+}W_{z+} + p_{z-}W_{z-}]$ being converted into a pure state.[117] If we regard the spin flip machine as part of the measuring device, we may say that this device performs a nonideal measurement of S_z, in which, irrespective of the result of the measurement and without loss of intensity of the beam, a pure polarized state $S_z = +\hbar/2$ is prepared; even if the incident beam is in a mixed state.[117]

The pure spin states physically determined by this apparatus comprise only ϕ_{z+}. This is an incomplete set. (The mathematical treatment of W would introduce the complete set $\phi_{z\pm}$.)

[117] This does not contradict the theorem that by the Schrödinger equation closed systems do not undergo transitions from mixed to pure states or *vice versa* as obviously the spins in the beam do not form a closed system. If we consider also the spatial part of the wave function, the flipping of the $S_z = -\hbar/2$ spin makes it impossible to use successive Stern–Gerlach apparatus (with fields directed ↓↑↑↓) to reunite the beams as in Wigner's thought experiment (see page 36), where it is the first of these four apparatus that "measured" S_z. After all spins have been turned into the $+z$-direction, we would have to use other means for reuniting the beams. We could never succeed *completely* in reuniting them, so that, mathematically speaking, we keep spatially a mixed state even if the spin state has become pure. However, we could possibly make the beams coinicide well enough to treat the result *as if* it were in one single pure state. (Remember that also in any *incident* beam there are always uncertainties that make its state a mixed one. Nevertheless, such initial states are often treated *as if* they were pure states, neglecting the uncertainties.)

Example 3.

We use the apparatus of Example 1 with one feature added. It is made to depend upon the thermal energy of some strategically positioned molecule μ in the apparatus, whether the spin turn mechanism shall be operative or not. The spins of the $S_z = -\hbar/2$ atoms will be turned to $S_x = +\hbar/2$ only if the energy ε of μ is more than some given amount ε_0. For $\varepsilon < \varepsilon_0$, the spins of the silver atoms remain unchanged.

If p_ε is the probability for $\varepsilon > \varepsilon_0$ in the mixed state of the apparatus, the final state now will have a density operator

$$W' = p_{z+} W_{z+} + p_{z-} p_\varepsilon W_{x+} + p_{z-} (1 - p_\varepsilon) W_{z-}.$$

The set of pure states determined physically by the apparatus here consists of ϕ_{z+}, ϕ_{x+}, and ϕ_{z-}, which form an overcomplete set. On the other hand, the mathematical treatment discussed above leads to a complete orthonormal set $\phi_{n\pm}$, where this time the angle θ of $\mathbf{n}$ is determined by

$$\tan\theta = p_{z-} p_\varepsilon / [p_{z+} - p_{z-} (1 - p_\varepsilon)].$$

This time it depends (through p_ε) upon the temperature of the apparatus, as well as upon the polarization of the incident beam.

In general, we shall use in the following unprimed names for the eigenfunctions between which the measurement of an observable differentiates, and from them we shall distinguish by a prime the noncomplete[118] nonorthogonal[118] set of pure states of which a mixture is formed by the (possibly nonideal) measuring apparatus together with gadgetry following behind it. We include with the gadgetry all further interactions that take place before the next measurement is performed upon the ensemble. Thus, there will be a prime placed upon the density matrix describing the ensemble incident upon the next measuring apparatus. This density matrix mathematically can also be interpreted as a mixture of a different set of pure states,

[118] By "non-" we mean here "not necessarily." So, "noncomplete" may be either complete, or uncomplete, where "uncomplete" is either incomplete (lacking), or overcomplete.

which we will distinguish from the primed states by using a tilde instead of a prime, and which form a complete orthonormal set.

In order to keep down the number of subscripts, we will often use lower-case italics without subscripts to indicate chosen eigenfunctions of an observable denoted by the corresponding capital italic. When we want to sum over all eigenfunctions, we change the italic into a corresponding Greek letter; so, when we want to sum over all eigenfunctions $u, x, y, z, a, b, c, d, e$, or f of the observables $U, X, Y, \ldots$, or F, we write for them $\upsilon, \xi, \eta, \zeta, \alpha, \beta, \gamma, \delta, \varepsilon$, or ϕ.

2.4. Alterations of state during and between measurements

Suppose some preceding apparatus prepares an incident mixed state in which $p_{z'}$ is the probability for the pure state z'. In general, the z' form a noncomplete[118] nonorthonormal[118] set. The ensemble E consists of subensembles $E_{z'}$ each in a pure state z'. The number of members of E present in the subensemble $E_{z'}$ is $N_{z'} = Np_{z'}$.

Next, we measure an observable A. Of the $N_{z'}$ members of $E_{z'}$, the number for which the instrument will react as if the incident state were the pure state a which is an eigenfunction of A (briefly: "for which a is measured") is given by

$$N_{z'a} = N_{z'}\,(z', a)^2. \tag{1}$$

Here we have introduced the notation

$$(z', a)^2 \equiv |\langle z' | a \rangle|^2 = (a, z')^2. \tag{2}$$

Division of (1) by N yields

$$p_{z'a} = p_{z'}\,(z', a)^2, \tag{3}$$

and summation over z' yields

$$p_a = \sum p_{\zeta'}\,(\zeta', a)^2. \tag{4}$$

While we use small letters p for the probabilities in the ensemble E on which all measurements are made, we denote by capital letters P the probabilities in subensembles of E, indicating the subensemble by superscripts on P. For instance,

$$P_a^{z'} \equiv N_{z'a}/N_{z'} = p_{z'a}/p_{z'} = (z', a)^2 \tag{5}$$

is the probability for finding a particular measurement result among the $N_{z'}$ members of the subensemble $E_{z'}$.

The result of measurement together with action of subsequent gadgetry is a mixed state in which there are probabilities $p_{a'}$ for physically determined pure states a' which form a noncomplete non-orthogonal system.[118] The $p_{a'}$ depend linearly upon the p_a, so that we may write

$$p_{a'} = \sum p_\alpha t_{\alpha a'}. \tag{6}$$

Here, $t_{aa'}$ is the probability of finding the final state a' in the subensemble E_a:

$$t_{aa'} = P_{a'}^a, \qquad N_{aa'} = N_a t_{aa'}, \tag{7}$$

and Eq. (6) follows from the second Eq. (7) by summing over a and dividing by N.

In E_a, the sum of the probabilities for the various a' should be unity:

$$\sum_{\alpha'} t_{a\alpha'} = 1. \tag{8}$$

However, in general,

$$\sum_{\alpha} t_{\alpha a'} \neq 1, \tag{9}$$

because, even if all measuring results a are found with equal probabilities [so that all p_a in (6) are equal among each other, and therefore each $= 1/n$, if n is the dimensionality of Hilbert space for the objects of our measurement], yet the outcome of the action of the gadgetry (as in Example 2) may show a preference for some particular final state a' [so that $p_{a'} \neq 1/n$ and n times Eq. (6) gives inequality (9)].

If the instruments are large compared to the objects **O** of the measurements,[115] the probability of a final state a' is the same in each of the subensembles $E_{z'a}$ as it was in E_a. Then,

$$N_{z'aa'} = N_{z'a} t_{aa'}. \tag{10}$$

The number of members in the subensemble $E_{z'aa'}$ for which a following measurement of B gives the result b is now

$$N_{z'aa'b} = N_{z'aa'}(a', b)^2 = N_{z'}(z', a)^2 t_{aa'}(a', b)^2. \tag{11}$$

Summing over all a and a' gives

$$N_{z'b} = \sum N_{z'\alpha\alpha'b} = N_{z'} \sum (z', \alpha)^2 t_{\alpha\alpha'}(\alpha', b)^2. \tag{12}$$

Further summation over z' and division by N yields

$$p_b = \sum p_{\zeta'}(\zeta', \alpha)^2 \, t_{\alpha\alpha'}(\alpha', b)^2. \tag{13}$$

Summing (11) over z' and a, and dividing by N, we get

$$p_{a'b} = p_{a'}(a', b)^2. \tag{14}$$

Thence,

$$p_b = \sum p_{\alpha'}(\alpha', b)^2, \tag{15}$$

which yields (13) by virtue of (6) and (4).

Note that, for any state a',

$$\sum_{\beta} (a', \beta)^2 = 1, \tag{16}$$

because the states b form a complete orthonormal set. Since the latter cannot be said about the a', in general we have

$$\sum_{\alpha'} (\alpha', b)^2 \neq 1. \tag{17}$$

In terms of the density operator

$$W' = \sum |\alpha'\rangle \, p_{\alpha'} \langle\alpha'| \tag{18}$$

of the ensemble E before the measurement of B, the probabilities p_b are expressible as

$$\begin{aligned} p_b &= \text{Trace}\,\{W'|b\rangle\langle b|\} = \langle b|W'|b\rangle \\ &= \sum \langle b|\alpha'\rangle \, p_{\alpha'} \langle\alpha'|b\rangle. \end{aligned} \tag{19}$$

Let W' be diagonalized by the complete orthonormal set of functions $\tilde{a}$:

$$\langle\tilde{a}_i|W'|\tilde{a}_j\rangle \equiv \tilde{W}'_{ij} = \delta_{ij}\,\tilde{p}_i, \tag{20}$$

so that

$$W' = \sum |\tilde{\alpha}\rangle \tilde{p}_{\tilde{\alpha}} \langle \tilde{\alpha}|. \tag{21}$$

Thence, (19) yields

$$p_b = \sum \tilde{p}_{\tilde{\alpha}} (\tilde{\alpha}, b)^2. \tag{22}$$

By (20) with $i = j$, $\tilde{p}_{\tilde{\alpha}}$ in Eqs. (21)–(22) is given by

$$\tilde{p}_{\tilde{a}} = \langle \tilde{a} | W' | \tilde{a} \rangle = \sum p_{\alpha'} (\alpha', \tilde{a})^2. \tag{23}$$

From (21) we see that mathematically we could interpret the state of our ensemble before the measurement of B, not only as a mixture with probabilities $p_{a'}$ for a noncomplete set of pure states a', but also as a mixture with probabilities $\tilde{p}_{\tilde{a}}$ for a complete orthonormal set of states $\tilde{a}$ which are the eigenfunctions of some operator $\tilde{\mathcal{A}}$, or of any function of $\tilde{\mathcal{A}}$. In this interpretation, the equality

$$\sum_{\tilde{\alpha}} (\tilde{\alpha}, b)^2 = 1 \tag{24}$$

replaces the inequality (17).

Inserting (6) and (4) in (23), and the result of this in (22), we obtain

$$p_b = \sum p_{\zeta'} (\zeta', \alpha)^2 \, t_{\alpha\alpha'} (\alpha', \tilde{\alpha})^2 (\tilde{\alpha}, b)^2. \tag{25}$$

We may now put

$$\tilde{t}_{a\tilde{a}} \equiv \sum t_{a\alpha'} (\alpha', \tilde{a})^2, \tag{26}$$

and obtain

$$p_b = \sum p_{\zeta'} (\zeta', \alpha)^2 \, \tilde{t}_{\alpha\tilde{\alpha}} (\tilde{\alpha}, b)^2 \tag{27}$$

to replace (13). Formally we thus have completely eliminated the noncomplete set of states a' physically realized by the apparatus, and we could say that our instruments change any measured state a into a mixture with probabilities $\tilde{t}_{a\tilde{a}}$ for the eigenstates $\tilde{a}$ of the "observable" $\tilde{A}$. However, as discussed at the end of Example 1 in Sec. 2.3, this set of states $\tilde{a}$ is not determined by the instrument alone, but depends directly upon the "polarization" $\{p_{z'}\}$ of the incident beam.

Moreover,

$$\sum_{\tilde{\alpha}} \tilde{t}_{a\tilde{\alpha}} = \sum t_{a\alpha'} (\alpha', \tilde{\alpha})^2 = \sum_{\alpha'} t_{a\alpha'} = 1, \tag{28}$$

$$\sum_{\alpha} \tilde{t}_{\alpha\tilde{a}} = \sum t_{\alpha\alpha'} (\alpha', \tilde{a})^2 \neq \sum_{\alpha'} (\alpha', \tilde{a})^2 \neq 1, \tag{29}$$

so we cannot avoid the inequality (9) by using the states $\tilde{a}$ instead of the states a'.

By repeating the arguments that led from (1) to (11) and from (4) to (13), we find for a longer chain of measurements $A, B, \ldots, D, E,$ with intervening interactions,

$$N_{z'aa'bb'\ldots dd'e} = N_{z'} (z', a)^2 \, t_{aa'} (a', b)^2 \, t_{bb'} \ldots t_{dd'} (d', e)^2 \tag{30}$$

and

$$p_e = \sum p_{\zeta'} \, \sigma_{\zeta' e} \tag{31}$$

with

$$\sigma_{z'e} \equiv \sum (z', \alpha)^2 \, t_{\alpha\alpha'} (\alpha', \beta)^2 \, t_{\beta\beta'} \ldots t_{\delta\delta'} (\delta', e)^2. \tag{32}$$

By repeated use of equalities like (8) and (16) we find

$$\sum_{\varepsilon} \sigma_{z'\varepsilon} = 1, \tag{33}$$

while the inequalities (9) and (17) in general yield

$$\sum_{\zeta'} \sigma_{\zeta' e} \neq 1. \tag{34}$$

Summation of (30) over the intervening indices yields

$$N_{z'e} = N_{z'} \, \sigma_{z'e}. \tag{35}$$

2.5. Time reversal in quantum theory

In the above, we have described *predictive* quantum theory. According to (35), in the subensemble $E_{z'}$ prepared to be in the pure state z', the probability $P_e^{z'}$ that later measurement of the observable E will yield the result e is for a given sequence of intervening operations predictable by

$$P_e^{z'} \equiv N_{z'e}/N_{z'} = \sigma_{z'e}. \tag{36}$$

Various authors have speculated about reasons why quantum theory should have this asymmetry in the time direction. Some have claimed that the *reduction principle* (see Sec. 2.2) is the source of this asymmetry, as it would always lead from pure states "before", to mixed states "after." (That is, transitions of subensembles $E_{z'}$ from their initial pure state z' to the mixture of subensembles $E_{z'a}$ each in a pure state a, and thence to a mixture of subensembles $E_{z'a'}$ each in a pure state a'.) However, it is easy to construct counterexamples. For instance, consider the subensemble E_a or $E_{a'}$. Both are in a pure state at some time during or after the measurement, but the relations $N_a = \sum N_{\zeta' a} = \sum N_{\zeta'}(\zeta', a)^2$ [compare Eq. (4)] and $N_{a'} = \sum N_{\zeta' a'}$ with

$$N_{z'a'} = \sum N_{z'}(z', \alpha)^2\, t_{\alpha a'} \tag{37}$$

indicate how E_a and $E_{a'}$ can be split into subensembles according to possible results of *earlier* measurements, so that *before* the measurement of A they were mixtures of subensembles $E_{z'a}$ or $E_{z'a'}$ in various different initial pure states z'. Thus, the reduction principle allows us to define also subensembles that go from mixed states before, to pure states after. The only lack of time symmetry here lies in our inability to recognize at the time of preparation of states z', *which* are going to belong to which E_a or $E_{a'}$, while we *can* recognize to which $E_{z'}$ they belong. The asymmetry therefore is in this case not in the principle of reducibility, but in the time asymmetry of our memory and thinking.

This has suggested to others that the asymmetry in quantum theory would be brought about by some kind of a physical interaction between the objects and apparatus of measurements on the one hand, and the human mind on the other hand. In objective quantum theory,[112] however, such interactions (if any exist) are not part of the matter discussed.

ABL[111] have shown that, if we stick to sequences of ideal measurements, *a probability interpretation of quantum theory can be formulated in a time-symmetric way*. [See Eq. (49) below.] This formulation does not reiterate the argumentations of the theory of measurements given in Chapter 1, but uses directly the *conclusions* of that theory, as listed in Sec. 2.2.

That notwithstanding this possibility a certain amount of irreversibility definitely exists in the applicability of quantum theory to probability problems is easily shown.[111] (See our Sec. 2.6.) ABL bring this irreversibility into the theory by introducing a postulate according to which an *initial* probability distribution shall predict a unique *final* probability distribution *independent* of what might occur *after* this final state is reached.

While this postulate is quite satisfactory from the point of view that one likes to construct a theory that shows aspects of causality, the authors themselves expressed some unhappiness that it is not *a priori* clear why not also the time-reversed postulate could have been made. Therefore, in Sec. 2.8 we will suggest replacing this postulate by a claim [Eq. (52)] which really should have been a theorem if we would have had the perseverance to prove it mathematically. Providing only some kind of a plausibility argument, because of our inability to give a satisfactory proof, we are at the present forced to introduce also Eq. (52) as a postulate, but it has this one advantage over the postulate of ABL, that we can immediately show [in Eq. (57)] that in general the time-reversed of *our* postulate *is impossible*, without using the asymmetry of quantum theory that according to Sec. 2.9 is a *consequence* of our postulate.

The source of this irreversibility in Eq. (52) we find to lie in the irreversible distinction between the equalities (33) and the inequalities (34). The latter were based upon the inequalities (9) and (17). This shows the importance of our generalizations over the work of Aharonov, Bergmann, and Lebowitz.

There is no doubt about the validity of Eq. (49), which is the basic postulate of "time-symmetric" quantum theory. (In the conventional predictive theory it is valid as a theorem. See Sec. 2.7.) The problem lies in retrieving the predictive laws of quantum theory from the postulate (49) with a minimum of additional assumptions.

Once the predictive laws have been retrieved, we find that our assumption (52) is equivalent to assuming that any ensemble has a tendency to spend in the far future most of its time in a mixed state in which all quantum states are about equally probable.[119]

Consider an ideal measurement of A immediately followed by a measurement of B. In predictive quantum theory we say that ideal measurements of A *pre*pare states a which allow us to *pre*dict probability distributions for the outcome of the following measurement of B. In the time-reversed problem, we say that the later measurement of B *post*pares states b, and we ask whether these states b allow us to *post*dict probability distributions for the outcome of the preceding measurement of A. The change of the prefix *pre-* into *post-* indicates the reversal of time.

ABL use the word *retro*diction instead of *post*diction. Personally I somewhat prefer the expression "postparing a state" to the lengthier "retroparing a state." The use of "retro" (= in opposite direction) for "post" suggests a preference for the direction toward the future; otherwise, "pre" would be as much *retro* to "post," as "post" is *retro* to "pre." The use of "retro" exclusively for "post" presumes that the time symmetry of the theory is broken. This symmetry *is* broken by the inequalities (34) for nonideal measurements and gadgetry. This justifies the use of "Retrodiction" in the title of this chapter.

This time asymmetry and the fact that for arbitrary ensembles the predictive rules of quantum theory and corresponding postdictive rules are mutually exclusive (see Sec. 2.6) do not mean that retrodiction is altogether impossible. In Sec. 2.6 we will find special conditions under which retrodiction is allowed, and in Sec. 2.10 we will give an illustrative example of this case. Application of retro-

[119] This was already discussed by ABL.[111] Their arguments, however, are of too limited validity, as they did not consider nonideal measurements interspersed with gadgetry.

diction in cases where these special conditions are violated leads to nonsensical results. (See Appendix C.)

It is perhaps useful to place at once some practical restrictions upon retrodiction, for excluding mathematically feasible but physically meaningless examples of it.

When treating the determination of b in a measurement of B as a "postparation" of the state of the ensemble E_b preceding the time t_B of the measurement, no difficulties are created if the measurement of B is nonideal. If, however, the preceding measurement of A was nonideal, or if some gadgetry was involved in preparing the states a' after measurement of A, the question arises what observable could by these instruments be measured upon the ensembles E_b with postdictable probabilities of the possible results?

From Sec. 2.3 we know that mathematically we may describe the apparatus and gadgetry as a machine preparing pure states $\tilde{a}$ that are eigenfunctions of some "observable" $\tilde{A}$ (and of any function of $\tilde{A}$). Should we say that we can (under proper conditions) postdict the probabilities for obtaining these eigenstates of $\tilde{A}$ in the postpared ensembles E_b?

Since these eigenstates of $\tilde{A}$ were determined by Eq. (20), *i.e.* by diagonalization of the density operator

$$W' = \sum |\alpha'\rangle\, p_{\zeta'}\,(\zeta', \alpha)^2\, t_{\alpha\alpha'}\,\langle\alpha'|, \tag{38}$$

they would depend not only upon the apparatus, but also upon the probabilities $\{p_{z'}\}$ describing the incident beam.

The *main* objection against this interpretation of postdiction, however, is that any probabilities postdicted for the eigenstates $\tilde{a}$ would not be experimentally verifiable, because our apparatus does not possess a pointer of which the position would tell us which eigenstate of $\tilde{A}$ would be measured. The only pointer present is the one indicating the state a.[120]

[120] As well known from the work of Daneri *et al.*,[121] the pointer position in many cases is attained *after* the interaction of the apparatus with **O**, by an irreversible process in the ordinary time direction.

Therefore it is wise to restrict all applications of retrodiction at least to those special cases where $\tilde{A}$ happens to be identical with the observable A of which the state can be read from the pointer position. That is, we would prefer to restrict retrodiction to *ideal measurements immediately preceding postparation with no gadgetry intervening*, so that

$$t_{aa'} = \delta_{aa'}, \qquad \tilde{a} = a' = a, \qquad \tilde{A} = A. \tag{39}$$

Of course, we do not want to impose restrictions like (39) upon measurements preceding A or following B.

As we shall see in Sec. 2.6, *the conditions (39) are not at all sufficient for validating retrodiction.* An additional requirement for this is that *the initial state shall be garbled.* We shall discuss in the following what this means, and how it is achieved.

2.6. Derivation of postdictive probabilities from predictive probabilities

Still clinging to the predictive postulate Eq. (5), we may ask in the ensemble E_b for the postdictive probability $P_a^b \equiv N_{ab}/N_b$ of having obtained the result a by the preceding ideal measurement of A. We assume here the restrictions (39). The predictive postulate (5) in this case gives

$$P_b^a \equiv N_{ab}/N_a = (a, b)^2, \tag{40}$$

so that

$$P_a^b = (a, b)^2 \, N_a/N_b. \tag{41}$$

Quantum theory would be time-symmetric in its probability rules if instead of (41) we would find the postdictive rule

$$P_a^b = (b, a)^2 = (a, b)^2 \tag{42}$$

[121] A. Daneri, A. Loinger, and G. M. Prosperi, *Nuclear Phys.* **33,** 297 (1962) and *Nuovo Cimento,* **44B,** 119 (1966). See also L. Rosenfeld, *Progress in Theoretical Physics,* Supplement, Extra Number on Meson Theory, 222 (1965), for a clarifying summary of the first of these two papers.

obtained by time reversal from (40). However, the dependence of P_a^b upon a required by (42) is found from (41) only if the factor N_a/N_b is independent of a, that is, if

$$N_a = \text{ (the same for all } a) = N/n, \tag{43}$$

where n is the number of linearly independent quantum states for the objects **O**. While from (41) we see that (43) is a *necessary* condition for the simultaneous validity of the predictive rule (40) *and* the postdictive rule (42), Eq. (43) is also a *sufficient* condition. This is so because the validity of (43) and (40) makes $N_b = \sum N_{\alpha b} = \sum N_\alpha(\alpha, b)^2$ equal to

$$N_b = (N/n) \sum (\alpha, b)^2 = N/n = N_a, \tag{44}$$

where by (39) we escaped the inequality (17) and could, instead, use the sum $\sum (\alpha, b)^2 = 1$ over the complete orthonormal set $\{a\}$. Equation (42) now follows from (41) with (44).

We shall call a mixed state (uniformly) *garbled*, if it is a mixture of a complete orthonormal set of pure states all occurring with the same probability. As the n equal probabilities add up to 1, the density matrix for a garbled state is

$$W = \boldsymbol{I}/n, \tag{45}$$

where $\boldsymbol{I}$ is the unit matrix. Therefore, the necessary and sufficient condition for validity of the postdictive rule (42) in the case considered is that *the state of the* e n t i r e *ensemble E be uniformly garbled at the time t_A of the measurement of A.*

Note that by (44) the garbling of the state of E survives from the time t_A until the time t_B of the measurement of B, but that *in general it will be lost after t_B*, because for the next measurement C we find

$$N_c = \sum N_{\beta\beta' c} = \sum N_\beta \, t_{\beta\beta'} \, (\beta', c)^2 = (N/n) \sum t_{\beta\beta'} \, (\beta', c)^2 \neq N/n \tag{46}$$

on account of $\sum t_{\beta b'} \neq 1$ and in general $\sum (\beta', c)^2 \neq 1$. [See (9) and (17).]

Garbling of the state of E at the time t_A may easily be accomplished by purposely garbling it at the time $t_{Z'}$ of preparation of the incident beam:

$$\{z'\} = \text{a complete orthonormal set } \{z\}, \quad \text{with} \quad N_z = N/n. \qquad (47)$$

Then, $N_a = N/n$ follows from $N_{z'} = N_z = N/n$ in the same way as $N_b = N/n$ followed from $N_a = N/n$. [See Eq. (44).]

Obviously, (47) is a sufficient, but not a necessary condition for (43). If the initial states z' do *not* form a complete orthonormal system, the necessary condition for (43) is[122]

$$\sum p_{\zeta'} (\zeta', a)^2 = 1 \quad \text{for all } a. \qquad (48)$$

If the set $\{z'\}$ is *in*complete, in general there will not be enough adjustable parameters $p_{z'}$ for satisfying (48).

In Sec. 2.10 we shall discuss an example in which the conditions (47) are met.

2.7. Derivation of the time-symmetric probability postulate from the conventional predictive one

Consider again the ensemble E of N systems **O** on which a sequence of nonideal measurements ..., Z, A, B, ... is made, with gadgetry or other interactions interspersed. As before, the **O** are assumed before each new measurement to be separable from the preceding apparatus, although their character may have changed (see pp. 57-58).

In Sec. 2.4 we derived for this case, from the predictive probability postulate, the Eqs. (30) and (35) for $N_{z'aa'bb'\ldots dd'e}$ and for $N_{z'e}$ in terms of $N_{z'}$. From these equations we can eliminate $N_{z'}$. We obtain

$$\begin{aligned} P^{z'e}_{aa'b\ldots dd'} &\equiv N_{z'aa'b\ldots dd'e}/N_{z'e} \\ &= (z', a)^2\, t_{aa'}\, (a', b)^2 \ldots t_{dd'}\, (d', e)^2/\sigma_{z'e}. \end{aligned} \qquad (49)$$

In "time-symmetric" quantum theory, this result (49) is raised to the status of a postulate, while the predictive rule Eq. (1) or (5) loses this privilege. The main difference between (5) and (49) is

[122] Note that (48) removes the dependence of (38) upon the incident beam $\{p_{z'}\}$.

that in (5) we predict probabilities for ensembles with a definite *initial* pure state (there z'), while in (49) we postulate probabilities for ensembles with definite *initial and final* states (z' and e).

The above derivation shows that (49) follows as a theorem from the predictive postulate (5) or (1). On the other hand, the predictive probability rule (5) is *not* a direct consequence of the "time-symmetric" postulate (49) alone. It requires some argumentation or an additional postulate[119] to derive from (49) the conventional predictive probability rule (5) of quantum theory.

The reader should here be warned that, though the postulate (49) looks time-symmetric at first sight, it still contains a biased time direction. This is due to the lack of symmetry between the primed and the unprimed states which finds its expression in the contrast between the equalities (8), (16), (33) and the inequalities (9), (17), (34). The postulate (49), however, is time-symmetric when applied to sequences of ideal measurements without gadgetry in between as in the paper by ABL. The lack of time symmetry in the more general case is due not to the form of the postulate, but to the "time arrow" (instrinsic irreversibility) built into successful measurements as well as into ingenious gadgetry.[123]

2.8. Tendency toward asymptotic garbling of the quantum state

In Sec. 2.6 we found in predictive quantum theory that the postdictive probability rule (43) for ideal measurements is valid only if the initial state is garbled. For *macroscopic* systems, the garbled state is

[123] The irreversibility of the operation of most measuring apparatus was discussed by Daneri *et al.*[121] The irreversibility of the indelible recording of the result of the measurement (used in Chapter 1 to characterize the success of the measurement) lies in the existence of this record at all future times and at no past times. Ingenious gadgetry, by selection mechanisms which may be automatized but may also be operated by (intellect and) hand, are apt to produce "unlikely" initial conditions for a new measurement, from the more chaotic conditions that may have prevailed after the preceding measurement; thus, their operation displays the kind of time asymmetry which is described by (8)–(9).

used in statistical mechanics as the foundation on which are based the derivations of the distributions in the various ensembles used in that theory; for instance, the distribution for the canonical ensemble for a system **S** is derived by assuming a garbled state of the ensemble of macroscopic systems (**S** $\oplus$ **R**), where **R** is a heat reservoir at a given temperature. (See textbooks on Statistical Mechanics.)

We also found in Sec. 2.6 that, for an ensemble of objects on which nonideal measurements are made and gadgetry is applied, *the garbled state is not conserved.* [See Eq. (46).] This shows that the evolution matrix $U(t_2, t_1)$ in

$$W(t_2) = U(t_2, t_1)\, W(t_1)\, U(t_2, t_1)^{\dagger} \tag{50}$$

in general is not unitary.[124] This complication was absent in the work of ABL[111] because they confined their discussion to sequences of ideal measurements only. Therefore, in their discussions, if W ever somehow would become garbled as in (45), it would stay garbled forever after, showing that the garbled state then would be an equilibrium state for the ensemble. This made it plausible that under such restrictions W would have a tendency toward this garbled state[125] and would asymptotically reach it "in the infinite future." Applying the same idea to macroscopic systems, we would conclude that the laws of thermodynamics, derived in statistical mechanics from this garbled state, would become valid in that "infinite future." The fact that the laws of thermodynamics are found to be pretty well valid *now* would

[124] This should not surprise us, because we are dealing here with the evolution of systems that are not closed, and separation of the **O** from the apparatus after each measurement involves a *reduction* of the Hilbert space in which a Schrödinger equation could be used.

[125] ABL[111] give an example of unitary evolution matrices $\langle a_i \,|\, U(t_B, t_A) \,|\, a_j \rangle = \langle a_i \,|\, b_j \rangle$ of which successive application by (50) upon the density matrix would asymptotically lead to a garbled state already by a sequence of ideal measurements only. For this, it suffices to consider the case where, for any two consecutively measured observables A and B, all $(a_i, b_j)^2$ would have the same value ($= 1/n$), while the phases of the complex off-diagonal matrix elements $\langle a_i \,|\, b_j \rangle$ would be randomly distributed.

show that the "infinite future" lies closer at hand than its name would suggest, at least for macroscopic systems.[126]

By allowing gadgetry which by (46) may undo a garbled state already reached, we may at least slow down the process toward a garbled state. Yet we might believe that ultimately, after living beings will have stopped displaying their inventiveness in dealing with the objects **O** and their descendants, nature by its stochastic processes would lead at least asymptotically[127] toward the chaotic state (45) for most any collection E of objects **O** (or of their descendants). The additional arbitrariness of final states introduced by allowing *nonideal* measurements may even hasten this garbling process.

If we would believe the validity of predictive quantum theory with its Eq. (35) for predicting future distributions from given initial distributions, we might assume that *in principle* the conjecture just mentioned, about a tendency toward garbling in absence of inventive intervention, could be demonstrated mathematically by statistical considerations about the effect upon $\sigma_{z'e}$ [Eq. (32)] of all the factors $t_{bb'}$ or $(c', d)^2$ that could go into it in case of random interactions interpreted partially as "measurements" and partially as "gadgetry preparing for the next measurement." Even if garbling is not strictly conserved and therefore never can be reached with finality, we still should be able to prove that, among the final distributions $\{N_f\}$ that would follow by $N_f = \sum N_{\zeta'} \sigma_{\zeta' f}$ from some initial $\{N_{z'}\}$, one could always find one (for some measurement F far enough in the future) for which $N_f \approx N/n$ would hold with any desired precision, irrespective of what the initial distribution $\{N_{z'}\}$ might have been. What such a

[126] This fact may be due to the scarcity of Maxwell demons and their gadgetry. Correspondingly, for the ensembles of systems **O** considered here, reaching a garbled state may require first escaping the attention of scheming physicists and their gadgetry.

[127] Because, according to Sec. 2.9, (49) together with (52) would have the predictive rule as a consequence, assumption of *strict* validity of $\sigma_{z'f} = 1/n$ for the sets $\{f\}$ for *all* measurements F in the future would yield *strict* validity of $N_f = N/n$, which would contradict the lack of conservation of the garbled state in the predictive theory according to (46).

proof really would show is that (35), if valid, would yield

$$\sum N_{\zeta'}\sigma_{\zeta' f} = \sum N_{\zeta' f} = N_f \approx \text{(the same for all } f\text{)}$$
$$= N/n = (1/n)\sum N_{\zeta'} \qquad (51)$$

for all distributions $\{N_{z'}\}$, and therefore that, for long enough chains of random events entering into the definition (32) of the $\sigma_{z'f}$, with arbitrary precision

$$\sigma_{z'f} \approx 1/n \quad \text{for the complete set } \{f\},$$
$$\text{irrespective of the prepared state } z'. \qquad (52)$$

The validity of (52) then should follow as a statistical theorem from the definition (32) alone, and *this proof would remain valid if the predictive rule and Eqs. (35) and (51) would be dropped.* We therefore believe that Eq. (52) should be regarded as a *theorem* which in principle it should be possible to derive from the *definition* (32) by statistical considerations about the possible $t_{bb'}$ and $(b', c)^2$ descriptive of random interactions, *without* assuming the validity of the predictive rule in quantum theory. In as far as we are not yet able to give the proof of this theorem, we shall temporarily take (52) as an additional *postulate*, which will allow us to drop the postulate of ABL about the existence of a unique relation, causal from the past toward the future, between probability distributions for successive measurements.[128]

We repeat that we do not claim that (52) would be strictly valid for the sets $\{f\}$ for all measurements F of the far future.[127] We merely claim that far enough in the future, in the absence of inventive intervention, the $\sigma_{z'f}$ from one set $\{f\}$ to a next set will fluctuate around

[128] Instead of assuming (52) for the states f of *actual* future measurements F, ABL note that, after a measurement of B, there is a *hypothetical possibility of* making measurements F for which the $\sigma_{z'f}$ would satisfy (52), and that in that hypothetical case the predictive rule (40) would follow from (49) by the method discussed in our Sec. 2.9. Then invoking the postulate that, whatever measurements are *actually* made after B, there shall exist for the probability prediction P_b^a for B a formula that *does not depend* upon the choice of the measurements made after B, they conclude that the actual formula for P_b^a must be identical with the hypothetical one.

$\sigma_{z'f} = N/n$ in such a way that there exists some F for which $\sigma_{z'f} \approx N/n$ with any precision we desire.

More precisely, for the application of (52) in Sec. 2.9 it will be necessary to assume that, for any two *consecutive* sets of prepared states $\{a'\}$ and $\{b'\}$, *the formula (52) will be valid simultaneously for one and the same complete set $\{f\}$, with z' in (52) standing for any of the states from the set $\{a'\}$, as well as for any of the states from the set $\{b'\}$.*

We now must verify that the latter specific condition is compatible with the postulate (52) in its general form. Actually, we shall find that, if (52) is valid for z' identified with any of the states b' from the later set, then it is automatically valid also with z' replaced by any of the precedingly prepared states a'.

On account of (32), we have

$$\sigma_{a'f} = \sum (a', \beta)^2 \, t_{\beta\beta'} \, \sigma_{\beta' f}. \tag{53}$$

If we assume the validity of (52) for the states b', then

$$\sigma_{b'f} \approx (\text{the same for all } f) = 1/n \tag{54}$$

for all states b'. Then, (53) gives

$$\sigma_{a'f} \approx (1/n) \sum (a', \beta)^2 \, t_{\beta\beta'} = (1/n) \sum (a', \beta)^2 = 1/n \tag{55}$$

on account of (8) and (16). This is (52) for the states a'.

Irreversibility of our assumption (52)

We shall now show that there are cases for which the time-reversed of our assumptions is not possible. For this, let us consider an example in which conditions (47) are not met by the z'. For instance, let us consider the case that measurement A was preceded by the preparation of one single pure state $z' = z_0$ which is one of the eigenstates of the observable Z measured before. We assume that all objects **O** for which states z different from z_0 are measured will hit absorbers, so their descendants will be in a state which we shall denote by o' and for which $\langle o' | a \rangle = 0$ for all results a of the measurement A. We would have

$$t_{zz'} = \delta_{zz_0}\, \delta_{zz'} + (1-\delta_{zz_0})\, \delta_{o'z'},$$

for which $\sum t_{z\zeta'} = 1$ only if for $z \neq z_0$ the sum over z' includes the state o', but for which $\sum t_{z\zeta'}\ (\zeta',\ a)^2 = \delta_{zz_0}(z_0,\ a)^2$ irrespective of whether or not we include o', as long as we include z_0. In this case, $\sum t_{\zeta z'} = \delta_{z_0 z'}$ for $z' \neq o'$, and $\sum t_{\zeta o'} = n-1$. [Compare (9).]

The time-reversed of the problem discussed in (52) now considers a long chain of measurements $U, \ldots, X, Y$, interspersed with other interactions, preceding the measurements Z, A, and B. We assume that the measurement U was performed by nature, some time before the origin of life, upon the ensemble E of the ancestors of our systems **O**, and that the incident ensemble at that time was in a mixture of states t'. The time-reversed of (52) would be the claim that with a proper choice of U and $\{t'\}$ from among the many such measurements and preparations of state through which the ancestors of **O** have gone in those prehistoric times, one would *in principle* be able to prove that, for all eigenstates a of A, except for fluctuations from one set $\{t'\}$ in the far past to the next,

$$\sigma_{t'a} \approx (\text{the same for all } t') = 1/n', \tag{56a}$$

where the accuracy of this formula could be arbitrarily improved by appropriate choice of the set $\{t'\}$ far enough in the past.

However, if by some reasoning we *could* prove this, we could by that same reasoning also prove that

$$\sigma_{t'z} \approx (\text{the same for all } t') = 1/n' \tag{56z}$$

for any result z that measurement of Z might have. Then,

$$\sigma_{t'a} = \sum \sigma_{t'\zeta}\, t_{\zeta\zeta'}\ (\zeta',\ a)^2 = \sigma_{t'z_0}\ (z_0,\ a)^2 \approx (z_0,\ a)^2/n', \tag{57}$$

in disagreement with (56*a*).

One might perhaps object against this reasoning that the $\{t'\}$ in (56*z*) could be different from the $\{t'\}$ in (56*a*), but this objection is not valid: A change from one set of prehistorically prepared states $\{t'\}$ to other such sets might cause some fluctuations of the $\sigma_{t'a}$, but could not cause a multiplication of all of them by a fixed factor $(z_0, a)^2$ independent of the $\{t'\}$.

Therefore, the assumption that the statistical validity of (56*a*) could be a general consequence of the rules of quantum theory must be wrong. No similar objection can be raised against the assumption of general statistical validity of (52).

Another warning that (56*a*) is unlikely to be correct, is that its sum over all states t' would yield $\sum \sigma_{\tau' a} = n'/n' = 1$, in disagreement with the expected inequality (34).

This all confirms our belief that the inequalities (9) and (17) are descriptive of the reasons for the impossibility of assuming the validity of (56*a*), even if we believe the validity of (52).

Note that, *if* somehow (56*a*) would be valid for some $\{t'\}$ and for all possible results a of measurement of A, the probabilities P_a^b for these results could be postdicted in each subensemble E_b corresponding to a given result b of a measurement of B immediately following the measurement of A without gadgetry used in between. For ensuring the validity of the restrictions (39), the measurement of A should also be an ideal one. Then, from (56*a*) we would be able to derive

$$\sigma_{t'b} = \sum \sigma_{t'\alpha}\, t_{\alpha\alpha'}\, (\alpha', b)^2 \approx (1/n') \sum (\alpha, b)^2 = 1/n'. \qquad (56b)$$

The method used in the next section then will provide the postdictive relation $N_{a'b} = N_b\,(a', b)^2$ for this case, by inserting (56*a*–*b*) and (39) in Eq. (60) below, and by performing the summations.

2.9. Derivation of the predictive probability rule in "time-symmetric" quantum theory

We shall now derive the predictive rule $P_b^{a'} = (a', b)^2$ from the "time-symmetric" rule (49) together with the definition (32), the assumption (52), and the equalities (8).

We calculate $P_b^{a'} \equiv N_{a'b}/N_{a'}$ using

$$\begin{aligned} N_{a'b} &= \sum N_{a'b\beta'\gamma\gamma'\ldots\varepsilon\varepsilon'\phi} \\ &= \sum N_{a'\phi}\,(a', b)^2\, t_{b\beta'}\,(\beta', \gamma)^2\, t_{\gamma\gamma'}\ldots t_{\varepsilon\varepsilon'}\,(\varepsilon', \phi)^2/\sigma_{a'\phi} \\ &= (a', b)^2 \sum t_{b\beta'}\,\sigma_{\beta'\phi}\,N_{a'\phi}/\sigma_{a'\phi}. \end{aligned} \tag{58}$$

By proper choice of the final $\{f\}$, all $\sigma_{b'f}$ and by (55) also all $\sigma_{a'f}$ in (58) may be equated with arbitrary precision to $1/n$. Thence,

$$N_{a'b} = (a', b)^2 \sum t_{b\beta'}\,N_{a'\phi} = (a', b)^2\, N_{a'} \sum t_{b\beta'} = N_{a'}\,(a', b)^2. \tag{59}$$

This is the predictive rule.

From this, we can retrieve for time-symmetric quantum theory all results obtained in Secs. 2.4 through 2.6 from the predictive postulate. In particular Eqs. (35) and (51), which we dropped below Eq. (52), now regain their validity. Also, the reasoning of Sec. 2.6 becomes applicable again, which showed us that postdiction would be possible in special cases only, as it cannot be valid when insufficient garbling preceded the postdicted measurement.

It is easily verified that we cannot derive the postdictive rule (42) by time reversal of our above derivation of the predictive rule. We can time-reverse (58) to

$$\begin{aligned} N_{a'b} &= \sum N_{\tau'\upsilon\upsilon'\ldots\alpha a'b} \\ &= \sum N_{\tau'b}\,(\tau', \upsilon)^2\, t_{\upsilon\upsilon'}\ldots t_{\alpha a'}\,(a', b)^2 \,/\, \sigma_{\tau'b} \\ &= (a', b)^2 \sum \sigma_{\tau'\alpha}\, t_{\alpha a'}\, N_{\tau'b} \,/\, \sigma_{\tau'b} \end{aligned} \tag{60}$$

which further simplifies if (39) is satisfied, but that is where it stops Equation (57) shows us that, if one single pure state $z' = z_0$ has been prepared preceding the measurement of A, we cannot believe Eqs. (56*a*) and (56*b*) and therefore we cannot show equality of the sum in the last member of (60) to $\sum N_{\tau'b} = N_b$, even if $t_{aa'} = \delta_{aa'}$ by (39).

Again we see how impossibility of time reversal is caused by the possibility of preparing pure states like z_0, by shunting all other states from the measurement of Z to the "absorbed state" o'.

The lack of time symmetry even in a theory based upon the "time-symmetric" probability postulate (49) does not come as a surprise:

We warned the reader already in Sec. 2.7 that time-asymmetry was built-in, and the asymmetry in the sense of Eqs. (8)–(9), of the $t_{zz'}$ of the example discussed in Sec. 2.8, is just a symptomatic example of this.

As ABL did not consider cases in which our inequalities hold, they were forced to invert some of the reasoning. Instead of deriving the impossibility of general validity of retrodiction by showing the *a priori* irreversibility of their additional postulate, as we have tried to do for our additional postulate, they derived the irreversibility of their additional postulate from the fact that its time reversal would lead to general validity of the postdictive probability rule (42), of which they knew that in general it had to be wrong, on account of its incompatibility with the predictive rule for ensembles that [contrary to the requirement (43)] were not initially garbled.

2.10. An example of valid retrodiction

In this and the following sections, we will confine ourselves to ensembles subjected to sequences of consecutive ideal measurements only, as in the original paper of Aharonov, Bergmann, and Lebowitz.[111] We then may drop the primed states from our formulas, identifying them with the corresponding unprimed ones.

In the next section we will reword some of the conclusions implicit in the work of ABL about those cases in which retrodiction is possible. In their last chapter, however, where these authors correctly point out that postdiction contrary to prediction is *not always* possible, their wording might trick some readers into believing that they would claim that actually postdiction *never* would be possible. We refer here primarily to their discussion of a sequence of Stern–Gerlach experiments (in the lower-half right-hand column of page B1415 of their paper), in which they say that the probabilities of a preceding measurement could not be retrodicted because "surely these probabilities would not be independent of the state of polarization of our beam prior to the performance of the first experiment."

We shall clarify this case by making their example more specific. Consider a beam of silver atoms traversing a sequence of Stern–Gerlach experiments in which consecutively we make ideal measurements Z, A, B, C and record the results. In particular, let us choose $Z \equiv S_{x'}$, $A \equiv S_x$, $B \equiv S_z$, and $C \equiv S_{z'}$, where $S_{x'} \equiv (S_x+S_z)/\sqrt{2}$ and $S_{z'} \equiv (S_z-S_x)/\sqrt{2}$. In this case, depending upon spin up or down, the two possible values of $(z, a)^2$ and similarly of $(b, c)^2$ are $\cos^2 22\frac{1}{2}° = 0.854$ and $\sin^2 22\frac{1}{2}° = 0.146$. For $(a, b)^2$, they both are 0.500.

The argument of ABL that the postdiction rule (42) would not apply in this case boils down to the remark that, in our example, (42) would require that, for either choice of a and of b (*i.e.* for either values $\pm\frac{1}{2}\hbar$ of $A \equiv S_x$ and of $B \equiv S_z$), the postdictive probability P^b_a should have the value 0.500 independent of the polarization of the beam incident upon A (that is, *independent of* z), while actually in this case $P^{zb}_a = 0.854$ or 0.146 depending upon whether the angle between $Z \equiv S_{x'}$ and $A \equiv S_x$ is 45° or 135°.

However, there is no contradiction at all between these values of the "interpolative" probabilities P^{zb}_a for a in the subensembles E_{zb} of E_z, and the postdicted value 0.500 of the postdictive probabilities P^b_a for a in each of the two subensembles E_b of E, where E is the initially garbled ensemble in which $N_z = \frac{1}{2}N$ for either choice of z. In each of these subensembles E_b (initially garbled according to $N_{zb} = \frac{1}{2}N_b$, so that $P^b_z = \frac{1}{2}$ for both values of z), the above-mentioned values of P^{zb}_a give[129]

$$P^b_a = \sum P^b_\zeta P^{\zeta b}_a = \tfrac{1}{2}\times 0.854+\tfrac{1}{2}\times 0.146 = 0.500. \qquad (61)$$

In other words, it is true that Eq. (42) does *not* postdict the probabilities P^{zb}_a of a for the subensembles E_{zb} of given b in either of the two ensembles E_z (one with $Z \equiv S_{x'} = +\hbar/2$, and one with $Z \equiv S_{x'} = -\hbar/2$), because these E_z are initially polarized. Yet, Eq. (42) *is* applicable for postdicting P^b_a in the ensemble E initially prepared in a garbled state, with equal probabilities for all n states (in this case two states) of a complete orthonormal set $\{z\}$. This is in full agreement with (and

[129] We use here $P^b_a = \sum N_{\zeta ab}/N_b = \sum N_{\zeta b}P^{\zeta b}_a/N_b = \sum P^b_\zeta P^{\zeta b}_a$.

is a good illustration of) the result we obtained around Eq. (47) in Sec. 2.6.

Our example seems to contradict the claim of ABL (on lines 35 and 36 of their page B1416) that "there simply is no way of avoiding preselection completely." I assume they try to say that, if one attempts to verify experimentally the postdictive probability rule (42) under its necessary condition of validity (43), it may be hard to garble the postdicted states *a* in the ensemble *E* while measuring the relative frequencies of the various postdicted events *a* in subensembles E_b corresponding to postselected final states *b*.

However, it is often relatively simple to garble the states *z* of the incident beam, as in the sufficient condition (47). In our example above, this may be achieved simply by using an unpolarized incident beam.

Moreover, the rules of quantum theory are rules about probabilities in well-defined ensembles,[112] and are not rules about the ease of experimental verification. If it would be hard with one single experimental arrangement to garble the initial states *z* (the "polarization of the beam") as in (47), we can always *combine* measurements made with *different* preselected initial states (ensembles E_z corresponding to different polarizations of the beam), and count in each case only those experiments in which the final event *b* happens to result. Thence we obtain the N_b and N_{ab} in an initially garbled ensemble *E* as the sums $\sum N_{\zeta b}$ and $\sum N_{\zeta ab}$ of counts in separate measurements, taking care that we use a complete orthonormal set of initial states *z*, and that we keep the N_z equal among each other.[130]

Anyhow, the possible occurrence of complications of this kind does not falsify the postdictive rule under the conditions for its validity. It would be very strange indeed, if under those conditions the postdictive rule would not hold, as we obtained it in Sec. 2.6 from the well-established predictive rule.

[130] The reason why this arrangement for an experiment on postdiction sounds so artificial lies partially in the lack of time symmetry in our *language*, as the latter mirrors the time asymmetry of our memory, as well as the distinction we make between "cause" and "result" on the basis of their time order.

2.11. Predictive and postdictive state vectors

Consider the ensemble E on which this time a sequence of ideal measurements $Y(1)$, $Z(2)$, $A(3)$, $A(4)$, $B(5)$, $B(6)$, $B(7)$, $C(8)$, $D(9)$, ... is made, where the numbers are used to indicate the order of sequence of the measurements. We assume that the observables measured are conserved between the measurements. For a specific example, consider again Stern–Gerlach measurements, so we may distinguish the two possible measurement results of any one by + and by −. Let $E_{3+,\,7-}$ mean the subensemble in which the results of measurements (3) and (7) were + and −, respectively. (States a_+ and b_-.)

As the reproducibility of ideal measurements guarantees that $E_{3+,\,4+} \equiv E_{3+} \equiv E_{4+}$ and that $E_{3+,\,4-}$ is empty, we say that measuring $A(3) = +$ *prepares* the state a_+ for the ensemble E_{3+}, which enables us to predict the result of measurement $A(4)$ on this subensemble. We could postpone the measurement $A(4)$ to any time preceding measurement $B(5)$. We therefore say that the state a_+ prevails in E_{3+} until the time t_5.

We shall now assume again, as in the preceding section, that the state of the *entire* ensemble E is uniformly garbled initially, and therefore in our special case is garbled for all times to come. Then, postdiction in E will be as valid as prediction.

Thus, a result $B(6) = -$ would not only *prepare* a state b_- for E_{6-} prevailing until t_8, but it will also *postpare* this state b_- for all times from t_4 on until t_6.

During the interval from t_5 till t_7, then, in the ensemble $E_{6-} \equiv E_{5-} \equiv E_{7-}$, the state b_- will be simultaneously the state prepared by $B(5)$ and the state postpared by $B(7)$, and it would seem to make some sense to say that the systems that are members of this ensemble during this time would each "have" the value $B = -$ for this observable.

On the other hand, during the interval from t_4 till t_5, in the subensemble $E_{4+,\,5-}$ obviously the prepared state is a_+ and the postpared state is b_-. That is, there are in this ensemble 100% probabilities for both $A = +$ and $B = -$, while $(a_+, b_-)^2$ is the probability of $B(5) = -$ in E_{4+}, and of $A(4) = +$ in E_{5-}.

The ensemble $E_{4+,5-}$ in our case is identical with the ensemble $E_{3+,7-}$. In the interval from t_3 till t_7, in this ensemble the state a_+ is predicted and b_- is postdicted. This prediction and postdiction is verified by the outcomes of the measurements $A(4)$, $B(5)$, and $B(6)$. However, obviously no general confirmation for all members of $E_{3+,7-}$ would be obtained if we would measure A, during the time interval from t_3 to t_7, *after* having made a measurement of B during that time interval. Therefore, there is a *statistically verifiable difference* between the correct statement that for $E_{3+,7-}$ a state a_+ was *prepared* for $t > t_3$ (until perturbation) and a state b_- was *postpared* for $t < t_7$ (from the last-preceding perturbation), and the "classical" (and *false*) statement that the state of the systems in $E_{3+,7-}$ would *be "a_+ and b_- simultaneously*," ignoring the need for a certain time order in the simultaneous verification of these states.

We conclude that it is *correct* to describe an ensemble like $E_{3+,7-}$ simultaneously by two different state vectors (one a predictive state vector and one a postdictive state vector) which are eigenfunctions of the predicted and the postdicted observables, which need not commute with each other; but, we should avoid losing the distinction between predicted and postdicted values of observables, and therefore we should *not* say that the systems in $E_{3+,7-}$ would simultaneously "have" $A = +$ and $B = -$.

2.12. The "practical" usefulness of retrodiction

Having established the possibility of retrodiction under proper conditions, we must admit that there exists one important reason for which one may want to ignore retrodiction in quantum theory. Human interest is centered more upon predicting the future than upon postdicting the past.[131] For this reason, questions involving retrodiction are seldom asked.

[131] The past can be "remembered," and probability considerations seem superfluous. About the future, calculating probabilities usually is the best we can do.

Most physicists (including Bohr) therefore customarily avoid talking about retrodiction at all. All talk about retrodiction looks like a waste of time. If we make the convention that by "state vector" or "density matrix" of an ensemble we *always* mean a *predictive* state vector or density matrix (and if we therefore *never* consider ensembles for which some future measurement result has already been selected), we can claim that (but for a phase factor) any objectively defined ensemble has only *one* (objectively determined) state vector or density matrix.

Under these circumstances, the conventional "Copenhagen rules" of quantum theory follow.[132] But then, of course, these rules should never be applied to retrodiction.

2.13. Misunderstandings in the literature

Misunderstandings caused by lack of distinction between prediction and postdiction have cropped up in the literature frequently. We will here indicate a few cases, referring the reader to some of the appendices for details.

Some of these misunderstandings are due to false philosophizing. Some people define "realism" as the doctrine that, for an individual "elementary" system occurring in nature, observables would "have" certain values, whether we know them or not.[133] This doctrine must

[132] We assume here again tacitly that the reduction problem has been settled. (See Chapter 1.)

[133] This particular brand of "realism," which in Sec. 1.4 we have denounced as being "*pseudo*realistic" if objective quantum theory correctly describes nature, may be subdivided into two types. Some people have tried to maintain that *every* observable simultaneously would "have" a certain value. Others, more modestly, assume that at any specific time each elementary system would "have" specific values merely for the maximal number of observables for which dispersionfree pure states of ensembles can be described by state vectors. This second type of (pseudo)realism is called *micro-objectivism* by d'Espagnat, who points out in sec. 16.2 of his 1971 book[134] that even this modest use of the (pseudo)realistic idea is contrary to the teachings of quantum theory.

[134] B. d'Espagnat, *Conceptual Foundations of Quantum Mechanics* (W. A. Benjamin, Menlo Park, 1971). Compare footnote 5 on our page 3.

"be" in direct disagreement with quantum theory, for, if it were valid, there could exist a (sub)ensemble of silver atoms that simultaneously would have definite values for all three components of its spin. Now, we can form a subensemble which is simultaneously in the predictive state $S_z = \hbar/2$ and in the postdictive state $S_x = \hbar/2$, but then it cannot either predictively or postdictively have a definite value of S_y. Therefore, quantum theorists would call this doctrine of "realism" *un*realistic (*i.e.* in contradiction with empirical knowledge).[135] We therefore called this doctrine "*pseudo*realism" in Sec. 1.4.

We purposely used here the example of the spin, where there are *three* mutually exclusive observables. Many authors, in connection with the question of (pseudo)realism, have tried to construct methods by which one could "know" simultaneously *two* noncommuting observables (x and p) for some given single physical system. Two examples of this, one appearing in a paper by Park and Margenau,[136] the other one in a paper by Ballentine,[13, 14] are discussed in Appendices A and B. Their reasonings are based upon a mixing of the ideas of prediction and of postdiction in definitions that are quantum-mechanically not unpermissible, but useless.

Worse is the confusion between prediction and postdiction in one interpretation of an unsatisfactory "proof" by which d'Espagnat in his recent book[134, 137] has tried to show that the state of an ensemble E of silver atoms, interacting with an apparatus for measuring S_z,

[135] While this kind of "realism" has no place in *quantum theory*, it could have a place in theories of elementary systems such as *hidden-variables theories*. Agreement between this kind of theories and the (pseudo)realistic philosophical doctrine, however, is for physicists irrelevant until experiments would show that these theories would describe nature better than quantum theory does. See Sec. 1.4.

[136] J. L. Park and H. Margenau, *Internat. J. of Theor. Phys.* **1**, 211 (1968).

[137] D'Espagnat's book[134] also expresses rather strange ideas about what would have been Bohr's point of view about the "time at which reduction takes place." According to d'Espagnat, a single physical system **O** according to Bohr would "acquire" the value A_n for the observable A measured at some time t_A upon **O**, and the time at which A would get this value would be the time at which the experimenter making the measurement at t_A would have first *planned* to make that measurement. (*Loc. cit.*, pages 173–175, 308, 340–342, 404.) In contrast, see our own Sec. 1.15.

after the measurement could not be a mixture of states with $S_z = +\hbar/2$ and with $S_z = -\hbar/2$, so that the reduction postulate would have to be wrong. This proof[138] may be interpreted as using postdiction in a case where postdiction is not allowed, and as erroneously assuming that postdiction starting from a predicted state should always yield the initial state on which the prediction was based. For details, see Appendix C.

[138] *Loc. cit.*,[134] pages 297–301 and 331–339.

CHAPTER 3

CONCLUDING REMARKS

3.1. Limits of validity of quantum theory

In the preceding two chapters we have discussed some of the features of quantum theory. We have not tried to settle the question to what extent this theory is "valid," that is, in agreement with the facts. This is something that only experimental results can settle. So far, the record of quantum theory seems very good.[26, 27] The recent results of Holt[28] which we mentioned on page 10, however, require further investigation.

What we have done, therefore, was mainly a discussion of the question to what extent the theory would seem usable from a theoretical point of view. Any theory of physics has its inherent limitations. Among macroscopic theories, Newton's theory loses its validity for large velocities; special relativity theory loses its strict (long-range) validity in gravitational fields; the general theory of relativity may be valid on a cosmological scale, but there is no *a priori* guarantee that it would not require amendment. On a microscopic scale, we replace the "classical" theories just enumerated, by quantum theory; it is natural to ask at what level should the rules of quantum theory be expected to break down.

The rules of quantum theory which we have discussed were the conventional ones, developed primarily for the quantum theory of electrons in nonrelativistic approximation. This Schrödinger theory has been extended. Dirac has extended it to a relativistic theory. For an understanding of the Pauli exclusion principle and its application to positons as well as negative electrons, one introduces the so-called "second quantization," and one starts using the methods of quantum field theory. The latter methods may also be used for the discussion of photons and other particles. While quantum electrodynamics has had great success, fundamental difficulties still remain. In the quantum theory of elementary particles, there are even more unsolved problems; in fact, hardly any problem has been completely satisfactorily solved as yet. Therefore, it is too early to say that the general principles of quantum theory which we have discussed in Chapters 1 and 2 would not need some amendments in the final theory of special-relativistic quantum theory of particles and of fields. Discussion of those problems lies beyond the scope of this booklet. Even more so the problems of general-relativistic quantum theory.

3.2. The part played by our mind

In order to come to some conclusions, therefore, we have strictly limited the field of our considerations. We have not tried to philosophize about the problems of interplay between our minds and atomistic nature. Therefore, in the discussion of the problem of reduction of states (essentially reduction of ensembles) after a measurement has taken place, we have simply stated that we may "choose" to make such a reduction (and in practice usually would be wise to make this choice when possible). This choice is an action of the mind; it may be induced by what we observe; but we have refrained from discussing what mental processes would lead to our choice. That is, we have not discussed, and do not want here to discuss, how we decide to *apply* quantum theory for the description of observations we make.

The problem of how a theory is applied is a *general* problem which is not specific to quantum theory. When an astronomer finds that a planet is one second of arc off the place on the sky where he previously by classical mechanics had calculated to expect it, he will simply readjust the "initial conditions" for the planet's motion to correct his misjudgement, and from there on will be using his new data until proven wrong another time. In atomic physics, because of the intrinsic unpredictability of individual systems it is much more common to find deviations of positions measured on individual systems, from average positions calculated from wave functions. Asked after a measurement on an individual elementary system to make predictions for the later behavior of the latter, the physicist will say that he cannot do this, but he *can* make *statistical* predictions for an ensemble "prepared by" (*i.e.* chosen according to) the observations he happened to have made on the individual system. Like the astronomer readjusted his initial data when after a measurement he found it expedient to do so, also the quantum theorist will after a measurement readjust his initial data; he will do so by choosing to discuss from then on a suitably elected new ensemble. This is "reduction." There is no reason to be more surprised about its use in quantum theory, than in astronomy. There is no more reason to discuss the workings of the mind and its consciousness in the one case, than in the other case. Conventionally, physicists leave such considerations of application of a theory outside the discussion of the theory itself. We have done so for quantum theory, and then find that Niels Bohr was right when he called the theory thus expounded an "objective" theory. (See Sec. 1.9.)

3.3. Reducibility after measurement

The main topic of discussion in Chapter 1 was whether the "rule **F5**," which allows reduction after a successful measurement, was *compatible* with the rule of time dependence of states by the Schrödinger equation. We have quoted authors who have claimed that these two rules would be intrinsically *incompatible*, because they could construe hypothetical cases where the two rules would require mutually contradictory density matrices. Obviously, quantum theory would not be blindly applicable in such cases. This means that these authors have found hypothetical limits of validity of the theory. The existence of such limits of validity may bother a philosopher who is trying to raise a law of physics to the status of a law of mathematics that would be valid unexceptionally on the basis of some axioms decreed, but it would not bother a physicist who realizes that *no* law of physics has absolutely unlimited validity. On the other hand, a physicist is concerned about finding where the limits of validity of his theory lie *in practice*. This was the subject of Chapter 1.

We mentioned, in the first place, that the measuring apparatus should be big enough so that it would not be appreciably affected by recoil caused by the measurement. (See pages 27–28.) In the second place, for arriving at a reducible state before reduction is performed, at least one of two possible conditions should be satisfied. The one sufficient possibility was erasure of off-diagonal elements of the density matrix in the representation diagonalizing the observable measured, by development of incoherence between states corresponding to different measurement results. The other possible condition was the making of a record that would not be destroyed by later measurements. (See Secs. 1.21–1.23.) In cases where neither condition would be fulfilled, application of rule **F5** would not be permissible. (See the last paragraph of Sec. 1.20.)

Several authors have claimed that the first method would be an impossibility, notwithstanding the fact that for instance Bohm in his 1951 textbook on quantum theory[61] presented explicit examples[87] of how this method could be used for actual measurements. The denial by these authors was based upon purely mathematical arguments. They claimed that the time dependence of the density matrix would be a unitary transformation.[50] (If it were, this incoherence procedure would indeed not be possible.) The reason for their assertion was that they considered only *closed* physical systems, for which the hamiltonian would be independent of external parameters.[139]

From a theoretical point of view, closed systems are neat, as they behave according to some clean rules. In practice, closed systems do not exist in nature. We should therefore develop our theories for sys-

[139] External *c*-number parameters on which the time integral of the hamiltonian may depend may describe macroscopic objects with which the quantum system interacts, or microscopic variations of the geometrical conditions under which the object **O** interacts with the apparatus **A**.

tems that are *nearly* closed and follow rules that are *nearly* clean. We should keep in mind that the slight deviations from clean rules that in practice are unavoidable are in most cases sufficient to cause the incoherence needed for making the final state after a measurement reducible.[140] (See page 20 and Appendix D.)

It has been asked: If a system is not closed, and therefore we describe it by an ensemble in which certain external parameters[139] may be stochastically distributed,[140] why don't we split that ensemble into subensembles each with definite values for these parameters? In each of these subensembles, incoherence might be avoided, and the final state after a measurement would remain irreducible in it.[141] In practice, of course, it usually would be hard to set up an experiment so that among a million performances of it we could pick out the subensemble of a thousand cases in which these (usually microscopic) external parameters would have a definite value. *If* we could, and we would follow up the measurement[141] by further experimentation on one such a subensemble, we *should* indeed find the interference phenomena characteristic of cases in which reduction after the first measurement is not allowed. This finding would not upset the principles of quantum theory. On the contrary, it would confirm them.

An example in which interference effects were caused by *fixing* an external parameter was discussed on page 35 where interference of light behind two slits illuminated through a slit of variable position in front was found to produce no interference fringes unless the position of the slit in front was fixed. We found in this case that the appearance of the fringe pattern on the third screen went hand in hand with a loss of success of the preceding measurement of the direction of the incident photon from the slit in front to the upper or to the lower slit

[140] When in the conventional (formal) theory of measurements of Secs. 1.17 *sqq.* we introduced a state $\psi^{\mathbf{OA}}$ for the **OA** system, we made an oversimplification, as ensembles of **OA** systems on which the predictions of the theory could be verified in the laboratory would always show fluctuations in the geometrical parameters describing the position of **O** with respect to **A**, and therefore should have been described as being in a mixed state, with the pure-state components of the ensemble leaving the measurement with phases incoherent among each other. (See footnote 87 on p. 36.)

[141] If the result of the "measurement" would not in any way be "recorded."

in the second screen. While Bohr, for explaining the absence of interference fringes in case of a variable position of the slit in front, utilized the incoherence of phases on the third screen caused by the variability of the parameter determining the position of the front slit, *we* used on page 34 for explanation the orthogonality of states corresponding to different measurement results together with the absence of transitions between different results during follow-up measurements. This second method of making the result of a successful measurement a reducible state is effective even more often than the first one. (For the majority of all measurements, *both* reasons for reducibility of the final state are valid.) Without a record, a measurement can hardly be called successful. The records are in *most* cases on a macroscopic scale (though exceptions to this exist), and such macroscopic records are sufficiently insensitive to the effects of later measurements, to validate our reasoning in Sec. 1.21.

3.4. The time arrow in quantum theory

In Chapter 2 we have discussed the lack of time symmetry in quantum theory. We found restrictions on the validity of postdiction of probabilities of measurement results of the past, that do not seem to exist for prediction of probabilities for measurement results of the future, notwithstanding the fact that formally the Schrödinger equation could be integrated from the future into the past as well as from the past into the future. This time asymmetry was recognized already by Aharonov, Bergmann, and Lebowitz.[111] After replacing the conventional predictive probability postulate of quantum theory by a time-symmetric probability postulate [a special case of our Eq. (49) of Chapter 2], they found that it had to be completed by a principle asymmetric in time before it would lead to the validity of the conventional predictive rules of quantum theory. (See pages 68 and 82.)

We generalized their methods, and used a different postulate for introducing the asymmetry in time. [See Eq. (52) and the discussion following it on pages 77–78.] We were able to show that the time-reversed of the latter postulate was not possible. We could do so because some lack of time symmetry was present already before this postulate was made. This asymmetry between future and past was formally expressed by our inequalities (9), (17), and (34), as opposed to the equalities (8), (16), and (33). Asymmetry in time enters the theory because it is intrinsic both in measurements and in the working of many ingenious gadgets.[123] The irreversibility of most measuring procedures had in the past already been stressed by Daneri *et al.*,[121] who pointed out that the process leading from the microscopic interaction of the object of measurement with the apparatus, to the macroscopic "pointer position" or the making of a record, usually is an irreversible thermodynamic or statistically ergodic process.

On account of this intrinsic asymmetry in time, postdiction in ensembles of measurements is allowed only under the specific condition that there is a "garbled" initial state. [See Eq. (43) in Sec. 2.6.] Postdiction, however, serves little purpose, and one might as well avoid it altogether. (See Sec. 2.12.)

3.5. Quantum theory and philosophy

People often have tried to adapt quantum theory to some particular type of philosophy which they prejudicially cherished, instead of adapting their philosophy to the facts of nature observed, and to quantum theory if the facts of nature fit that. In Chapter 1 we have tried to follow the latter policy, but we have not much elaborated upon the philosophical consequences.

D'Espagnat[4, 5] and several other authors have tried to approach this subject from the other side. They have considered a number of different philosophical systems, and have asked themselves whether or not quantum theory would fit in these systems. Unfortunately, in the course of this attempt, they occasionally have ascribed to quantum theory qualities which this theory does not possess, such as their claim that objective quantum theory would be self-contradictory because of a supposed impossibility of ensembles of systems to go from a pure state into a mixed state.

Here and in some of the following sections, we want to say a few words about this question of the relation between quantum theory and philosophy, but we want to keep this very brief.

One philosophical attitude that many physicists in the past have found useful for not creating clashes between physics and philosophy, is the **positivistic** attitude. However, taken to its extreme, this easily leads to a **solipsistic** point of view.[142] This point of view raises more questions than it solves. In particular, it leaves unanswered the question why I have the impression that others observe the same things that I observe.

Moreover, intemperate positivism is not conducive to the creation of physical theory. Physical theories very often start out from some model of nature that we make in our mind. Discussing such a model we tend to use a language as if nature is an object about which we can talk without descending to the level of the immediate sensations by which we observe nature.

This leads us automatically to a **realistic** point of view. That is, we are tempted to ascribe to nature reality independent of our own existence. This realistic point of view has the advantage that it now becomes obvious why different persons observe the same things: Objective reality outside us interacts with the minds of different people in similar ways.

From this point of view—in contrast to the solipsistic point of view—the objective reality of matter is postulated to be primary, and our consciousness of it is only secondary.

3.6. Our mind and its workings

Our mind may be regarded materialistically as the workings of a blob of neurons and other biological entities serving as a circuitry in which complicated patterns of chemical reactions and of electric currents take place. The latter are the materialistic counterparts of "thoughts," "feelings," and other reflexes. No wonder then that the mind dies when the body dies. Learning and memory become bio-

[142] See, for instance, E. P. Wigner, *The Monist*, **48**, 248 (1964).

logical processes, like the growing of contacts between parts of this complex system of specialized cells.

From the quantum-mechanical point of view, this mind acts in as indeterministic a way as the remainder of nature; but, from statistical mechanics in its thermodynamic limit we know that much of this indeterministic action looks macroscopically pretty deterministic. On the other hand, an indeterminate atomistic happening in this kind of a biological system may from time to time trigger processes that magnify its effect to the macroscopic level.

From this point of view, as well as from daily experience about the working of our senses, we find the influence of our surroundings upon our mind to be more direct than the influence of our mind upon our surroundings. The latter takes place mainly indirectly, by our mind commanding our muscles to perform acts that influence our surroundings. Acting upon our surroundings directly by brain waves is much more difficult.

In particular, I do not know how to use my mind for forcing a measurement to come to a definite conclusion identical with its result, unknown to me and possibly to anybody, but by some device recorded automatically. As this record may be of a macroscopic nature, it is contrary to my realistic point of view about the existence of this record, to believe that this record would take its definite contents at the time when it is read, rather than at the time when the device created it. This is my main objection against some of Wigner's ideas.[143] In most cases, it would seem to me, nature chooses the result of a measurement without the intervention of my mind, and then nature influences my mind to become conscious of whatever it decided to do.

3.7. Quantum theory and God

People are accustomed to assume that only "persons" can decide what to do. (In this sense, *some* "personality" may be ascribed even to animals.) Therefore, if I get the impression that nature itself

[143] See pages xiii–xv of the Preface, and see footnotes 40, 41, and 47 on pages 17–18.

makes the decisive choice what possibility to realize, where quantum theory says that more than one outcome is possible, then I am ascribing personality to nature, that is, to something that is always everywhere. Omnipresent eternal personality which is omnipotent in taking the decisions that are left undetermined by physical law is exactly what in the language of religion is called *God*.[144] We thus see how quantum theory requires the existence of God.[145] Of course, it does not ascribe to God defined in this way any of the specific additional qualities that the various existing religious doctrines ascribe to God. Acceptance of such doctrines is a matter of belief or faith.

3.8. Quantum theory and realism

We mentioned in Sec. 3.5 that physicists have a tendency to accept to a certain extent the doctrine of realism. This means that we like to assume that our mind exists inside this world, in contrast to the solipsistic idea that the world would exist within our mind.

On the other hand, quantum theory shows that the validity of realism as a philosophy in agreement with physical facts is severely limited, at least at the level where quantum theory is operative. We have pointed out in Secs. 1.4, 1.10, and 2.11, and it has been well explained by d'Espagnat and by others, that in general quantum theory does not allow us to assume that observables of individual elementary systems would "have" definite values irrespective of their measurement. We cannot ascribe a state vector to an individual system, either, without losing most of the meaning of a state vector. (See Sec. 1.8.)

If elementary systems do not "possess" quantitatively determinate properties, apparently God determines these properties as we measure them. We also observe the fact, unexplainable but experimentally

[144] We follow the custom of the English language, of using capitals for indicating the omnipotence of God.

[145] Einstein did not like this. See footnote 108 on page 50.

well established, that God in His[144, 146] decisions about the outcomes of our experiments shows habits so regular that we can express them in the form of statistical laws of nature. Since the macroscopic behavior of nature may be thought as determined by the behavior of its zillions of atomistic constituents, these statistical laws of nature with great precision lead to apparently deterministic laws that seem to govern nature macroscopically. Especially during the nineteenth and the early twentieth century, this apparent determinism in macroscopic nature has hidden God and His personal influence on the universe from the eyes of many outstanding scientists.

At our present level of knowledge, it seems as if a few of the deterministic laws of nature would extend their validity from the macroscopic into the microscopic domain. I refer here to certain conservation laws. It is still believed that the charge of an electron can be predicted with certainty. It is believed that there exist ideal measurements which would allow us to predict an identical outcome for an identical follow-up measurement. If these beliefs are justified, this would show the admissibility of a *limited* amount of the "realistic" attitude about determinate properties of matter which *in general* we had to reject. Any generalizing of this realism in the writings of various philosophers appears, however, to be contradictory to the physical facts, as explained in Chapters 1 and 2.

This shows us the danger of defining realism as some *philosophical principle*, and then trying to force this definition upon physical theory. From the point of view of the physicist, a more satisfactory procedure is to use the laws of physics for figuring out under given circumstances how much "reality" to expect, and to avoid generalization of ideas of realism to other cases where the circumstances may be different.

[146] We follow here the custom of the English language, of expressing the Personality of God by using for Him the human-male pronoun. The reader should not misinterpret this. We regard God as a Power, and not as a biologically-male human being.

Physicists and philosophers

The above is just one example of the difference in attitude between a typical physicist and a typical philosopher. The philosopher hopes to formulate ideas of general validity that he hopes to be applicable to a field of study often much wider than the limited scope of one of the physical sciences. He then tries to derive conclusions from these ideas. In this procedure, the philosopher is closer to the mathematician than to the physicist.

The physicist is more pragmatical. He collects experimental data, and makes theories that within the limits of possible errors will describe these data; or, he makes physical theories, and then performs experiments to confirm within the limits of possible errors the validity of these theories. Quantum mechanics with its probability interpretation is just one of these theories.

When the physicist observes a philosopher at work comparing how various philosophical ideas jibe with quantum theory or with some other physical theory which the physicist considers to be valid, the physicist tends to regard this as a game, just like he considers the inventing of all kinds of esoteric geometries a game. When such a game succeeds—when a geometry is invented that explains geometrical observations in nature, or if a philosophical system is invented that fits existing theory—the physicist will gratefully accept and use the invention. Sometimes, however, the philosopher ends up with contradictions that do not seem to have a solution. Such a negative result still may be a mental achievement of the first rank, and it may be useful as a warning what philosophical ideas to avoid, though physicists may tend to regard such work as a lost effort.

What observables can "have" a value?

While elementary systems according to quantum theory do not "have" definite values for their observables, they still can be "*prepared tio have*" values (see Sec. 1.10), as they will be when they are predictvely in the corresponding *eigenstate* (see Secs. 2.11 and 1.8).

Where in Sec. 1.21 we admitted *quantum-mechanical* systems for *recording results of measurements* (compare footnote 90), the record of a preceding measurement would be a preparation of a definite eigenstate of the memory. For instance, the memory could be a crystal in which each molecule can have "spin up" or "spin down" in an external field. Specification of the eigenstate characterized by the spins of all its molecules could be regarded as a "record."

If, however, we want to be able to *read* and *know* the result of a measurement, we need for that a *macroscopic* record. According to *classical* physics, observables on *classical* physical systems *can* just plainly "*have*" definite values. Since classical physics gives a satisfactory description of most macroscopic objects, we may *talk in classical language about outcomes of measurements read from classical apparatus.*

Summarizing: We may say that *classical* systems may "have values" for all of their classical observables. Therefore, outcomes of measurements may have values. Correspondingly, quantum systems may have *postdictive* values for one set of commutative variables, and they may have *predictive* values for another (possibly but not necessarily the same) set of commutative variables. Since, however, the sets of predictively and of postdictively determined observables need not be identical, we should *avoid* simply saying that observables would *have values* on elementary systems.

APPENDICES

APPENDIX A

POSITION AND MOMENTUM ACCORDING TO PARK AND MARGENAU

We shall criticize here a method often suggested, and recently discussed by Park and Margenau,[136] by which one supposedly can find simultaneously the position x and the momentum p of a particle with arbitrary precision. For simplicity we consider the one-dimensional case. For allowing a confrontation of this method with the rules of quantum theory, we shall here apply it to an ensemble instead of to a single system.

We prepare an ensemble of which it is known with arbitrarily good precision at what times t_1 the particles pass the point x_1. From this ensemble, we pick the subensemble of particles that with similarly good precision pass the points $x_2 = x_1+\Delta x$ (with given Δx) at times $t_2 = t_1+\Delta t$ (with given Δt). Then, calculating each particle's momentum during the time interval $\{t_1, t_2\}$ by $p = m\Delta x/\Delta t$, we will know not only x_1 and x_2, but also p with arbitrary precision.

To start with, note that x_1 here does not commute with x_2. The operator $x(t)$ depends upon time by $x(t) = T(t)^{-1}\cdot x\cdot T(t)$ with $T(t) = \exp(tp^2/2mi\hbar)$. Therefore,

$$x\cdot T(t) = T(t)\cdot x + i\hbar\frac{\partial T(t)}{\partial p} = T(t)\cdot\left[x+\frac{pt}{m}\right],$$

so that $x(t) = x+(pt/m)$, which on the one hand justifies calculating p by $p = m\,\Delta x/\Delta t$, but which at the same time shows that there does not exist an ensemble which is simultaneously in predictive eigenstates of $x(t_1)$ and of $x(t_2)$.

This, of course, does not mean that one could not measure x_2 with arbitrary precision at t_2, after measuring x_1 with arbitrary precision at t_1 (according to the *nonrelativistic* quantum theory here under discussion). This is just as possible as it is to measure p itself at t_2 after measuring x at t_1. During the time between t_1 and t_2, then, the subensemble corresponding to precise values of both x_1 and x_2 is simultaneously in a predictive eigenstate of $x(t_1)$ and in a postdictive eigenstate of $x(t_2)$. Therefore it is somewhat misleading to say that we would "simultaneously know" the values of $x(t_1)$ and $x(t_2)$ which the systems in our subensemble would "have" during this interval. The usefulness of this knowledge of x_1 and x_2 would be no more[147] than the usefulness of the "knowledge" we would have of the states a and b in the ensemble E_{ab} between times t_A and t_B in Secs. 2.10 and 2.11. See, in particular, the last two paragraphs of Sec. 2.11. We saw there that this kind of simultaneous "knowledge" of two observables, A and B, is not at all contradictory to the principles of quantum theory, even if A and B do not commute. Since part of this "knowledge" is postdictive, however, it is "useless" in the sense of Sec. 2.12.

For given predictive value of $x(t_1)$ and postdictive value of $x(t_2)$ we can, as an exercise in arithmetic or as a historical fantasy, calculate for the ensemble $E_{x_1 x_2}$ the quantity $p = m\,\Delta x/\Delta t$; but we must keep in mind that this "knowledge" of p is neither predictive nor postdictive,

[147] Actually, in the interval between t_1 and t_2, the value of this knowledge would in the present case be *less*. In Sec. 2.11, where A and B were conserved quantities, we could simultaneously predict the result of measurement of A and postdict the result of measurement of B, at times t_a and t_b, at least as long as $t_A < t_a < t_b < t_B$. Since x is not conserved, knowledge of x_1 or of x_2 alone certainly would not make x predictable at any time between x_1 and x_2. The value of x measured between t_1 and t_2 cannot be determined even from *simultaneous* knowledge of x_1 and x_2, as in the ensemble $E_{x_1 x_2}$ at time t values for x different from $[x_1(t_2-t)+x_2(t-t_1)]/(t_2-t_1)$ would be found, for particles which at this intermediate measurement of x would pick up the right amount of recoil for reaching x_2 at time t_2.

as it combines in one formula a predictively known observable and a postdictively known observable, and the former is postdictively unknown, while the latter is predictively unknown.[148]

Further, the claim of Park and Margenau that x and $p = m\,\Delta x/\Delta t$ would commute[149] is simply wrong. Their error here is caused by their treating x_1 in $\Delta x \equiv x_2 - x_1$ as if it were a c-number, and their treating $x_2 = T(t_2)^{-1} \cdot x \cdot T(t_2)$ as if $T(t)$ would commute with x.

APPENDIX B

SIMULTANEOUS MEASUREMENT OF POSITION AND MOMENTUM ACCORDING TO BALLENTINE

In his paper on the statistical interpretation of quantum mechanics,[13, 14] Ballentine early in his chap. 3 points out that, in any ensemble E, the dispersions Δp and Δq (of which the product according to the indeterminacy principle is to be larger than $\hbar/2$) are root-mean-square deviations obtained in long sequences of measurements of p and q in which the p may be measured on the systems of one representative part of E, and the q on *different* systems from *another* part of E. From this it is clear that any uncertainty Δp found in the p is not caused by a measurement of q on the same system.

[148] If we would make ideal measurements of the momentum p on the ensemble $E_{x_1 x_2}$ between t_1 and t_2, we would find the subensemble $E_{x_1 p x_2}$ to be nonempty for p values different from $m\,\Delta x/\Delta t$, for the simple reason that this subensemble of $E_{x_1 x_2}$ would also be a (clearly nonempty) subensemble of the ensemble $E_{x_1 p}$.

[149] Reference 136, page 244.

On the other hand, it is also clear that, in an ensemble *prepared* to be in a state of rather well defined q, the conjugate p will be rather undetermined. This occurs when measurements of p *are* made on the same systems on which first an ideal and *selective* measurement of q was made, where by "selective" we mean elimination of all systems on which the measurement of q did not result in some particular chosen value for q. When Heisenberg and others claim that a measurement of q would *cause* an indeterminacy in p, obviously they refer to this kind of selective ideal measurements of q, and to measurements of p made on *the same* systems.

Ballentine on his page 366 rightly warns that this is a poor usage of the words "measurement of q," for what should have been called "preparation of an ensemble in a state of well-determined q." Properly, a measurement determines a state that *was* (and that need no longer be, if the measurement is not an ideal one). That is, a measurement, in the common sense of the word, is not a preparation of state, but is a *post*paration of state. On the other hand, this misuse of the word "measurement" for "preparation" is so widespread in the quantum-mechanical literature, that we have to reckon with it, so that we simply must be cautious not to interpret this word literally as meaning "postparation," wherever an author obviously meant (but failed) to say "preparation." Luckily some authors warn their readers by saying that they talk about *ideal* measurements. This gives away that they are more interested in what happens *after* the measurement, than in what preceded it.

While, up to this point, Ballentine is right in warning that, in an experimental *verification* of the indeterminacy principle, one should *avoid* considering measurements of p and of q made upon the same system, and while it goes without need for discussion that, *when* measurements of q and of p are made *consecutively* upon one and the same system, *both* measurements on each individual system may be made sharply, Ballentine goes too far when he claims that the uncertainty principle would not preclude a *simultaneous* sharp measurement of *both* q and p on one and the same system. That is, in our language, he claims that he can *post*pare a state in which both q and p would be known simultaneously. (He agrees, on page 366, that he cannot *prepare* a state in which q and p are known simultaneously.)

Let us see what is wrong with the alleged example of simultaneous measurement of y and of p_y constructed by Ballentine.[150]

Preceding this measurement, a beam of particles is prepared to have a given kinetic energy K. (Compare the preparatory measurement of Z in Sec. 2.10.) Next, a "measurement A" is made, consisting here of letting the incident particles traverse or not traverse a pinhole P_1 at position $\mathbf{x}_1$ in a screen. Here, the particles that pass through are elastically diffracted, so that they move in all directions with a given momentum $|p| = \hbar k = \sqrt{2mK}$. The predictive state $\boldsymbol{a}$ behind the hole may be described by a predictive state vector, something like[151] $\psi_1 = C_1 r_1^{-1} \exp(ikr_1)$, where $r_1 = |\mathbf{x}-\mathbf{x}_1|$. More precisely, ψ_1 is the state of the ensemble E_{za}, as it contains information not only on $\mathbf{x}_1$, but also on K. That the value of K is not lost at $\mathbf{x}_1$ may be checked experimentally by making measurements of K behind the screen on some arbitrarily picked subensemble of E_{za} representative of the whole ensemble.

Ballentine now considers on the ensemble E_{za} a follow-up "measurement B" consisting of letting the particles traverse or not traverse a pinhole P_2 at $\mathbf{x}_2$ in a second screen. Let E_{zab} be the subensemble that passes through P_2. Behind this second screen we may verify again that K was conserved. Assume this is done, successfully for all members of E_{zab}, by a "measurement C." Let the state $\boldsymbol{c}$ postdict the original value of K. We may then identify the ensemble E_{zab} with E_{zabc}.

We could also imagine doing measurements B and C on an *initially garbled* ensemble E (particles coming not just from P_1, but from anywhere). The subensemble of E, of particles passing through P_2 and subsequently having the kinetic energy K, we shall call E_{bc}. Its postdictive state vector before reaching P_2 is something like[151] $\psi_2 = C_2 r_2^{-1} \exp(-ikr_2)$, with $r_2 = |\mathbf{x}-\mathbf{x}_2|$.

[150] Reference 13, page 365 and fig. 3.

[151] We give this formula merely as an illustration: It does not correctly describe the diffraction in the pinhole, on which we gave insufficient information.

The ensemble E_{zabc} considered by Ballentine is a subensemble of E_{za} as well as of E_{bc}. In the time interval $\{t_1, t_2\}$, it is in a predictive state ψ_1 giving information on $\mathbf{x}_1$ and on $K = \hbar^2 k^2/2m$, and it is in a postdictive state ψ_2 giving information on $\mathbf{x}_2$ and K.

Postdictively, there is no information whatsoever on $\mathbf{x}_1$. Nevertheless, Ballentine claims that by measuring accurately the y-components of $\mathbf{x}_1$ and $\mathbf{x}_2$, he can get accurate postdictive information on the value of p_y preceding passage through P_2, from the y-component of

$$\mathbf{p} = \hbar k \mathbf{n} \quad \text{with} \quad \mathbf{n} = (\mathbf{x}_2 - \mathbf{x}_1)/|\mathbf{x}_2 - \mathbf{x}_1|. \tag{B.1}$$

Obviously, neither ψ_1 nor ψ_2 satisfies $(\hbar/i)\nabla\psi = \hbar k \mathbf{n}\psi$, because ψ_2 lacks the information on $\mathbf{x}_1$ needed in $\mathbf{n}$, and ψ_1 lacks the information on $\mathbf{x}_2$.

We recognize some similarity of Ballentine's claim Eq. (B.1), to Park and Margenau's claim that $\mathbf{p} = m(\mathbf{x}_2 - \mathbf{x}_1)/(t_2 - t_1)$ for particles passing through both P_1 and P_2. One difference is that, in addition to $\mathbf{x}_1$ and $\mathbf{x}_2$, Ballentine uses information on K instead of on $(t_2 - t_1)$.

We have shown in Appendix A that the formula of Park and Margenau by itself can be justified, by reading $\mathbf{x}_1$ and $\mathbf{x}_2$ as $\mathbf{x}(t_1)$ and $\mathbf{x}(t_2)$, and by using the time-dependent operators $\mathbf{x}(t)$ of the Heisenberg representation. If the expression for $\mathbf{n}$ in Eq. (B.1) is treated in the same way, we may interpret Eq. (B.1) as $\mathbf{p} = p\,\mathbf{p}/p$. Our objection is not against these equations themselves, but against their misuse.

As Ballentine inserts in (B.1) for $\mathbf{x}_1$ a value which for E_{zab} he knows predictively, and for $\mathbf{x}_2$ he inserts a value which he knows postdictively, he obtains for $\mathbf{p}$ a value which for the ensemble E_{zab} he knows neither postdictively, nor predictively. Therefore, he has no right to say that on the members of the ensemble E_{zab} he would have *measured* the value of the quantity appearing in the left-hand member of (B.1), that is, the momentum vector or its components. He simply has *defined* by it some quantity $\mathbf{p}$ of unknown physical meaning. As in the case of Park and Margenau (see the second last paragraph of Appendix A), we may regard this as either an exercise in arithmetic, or a historical fantasy based upon the prejudices of classical physics. We cannot expect this definition of $\mathbf{p}$ to tell us any experimentally verifiable facts about the ensemble E_{zab}.

If Ballentine by his "measurement B" really would have completed a *measurement* of $\mathbf{p}$ or p_y, then an ideal measurement of it *preceding* the particle's passing through P_2 should give an identical result without disturbing the state of the ensemble. In reality, of course, such a measurement would turn up also particles with *different* values of $\mathbf{p}$ and p_y, that nevertheless came from P_1 and that would traverse P_2 as well. The interaction of these particles with the measuring apparatus would see to that.

Experimentation would not even confirm the rectilinearity of the trajectory from P_1 to P_2. If we place a third screen with pinhole P_3 between the screens with pinholes P_1 and P_2, for making a "measurement D" of $\mathbf{x}_3$ between the measurements A and B, then in case of linearity of trajectory all of the ensemble E_{ab} should pass through P_3 when P_1, P_3, and P_2 would be lined up, and nothing should pass through when they would *not* be lined up. Instead, due to the elastic diffraction at P_3, we would find only *part* of E_{ab} to pass through P_3 when it is lined up, while we would find E_{adb} to be nonempty when P_3 lies *off* the line P_1P_2.

While it is easy to find an excuse for such zigzags by saying that the measurement of $\mathbf{x}_3$ perturbed the momentum of the particles on their ways from P_1 to P_2, we see that statements that measurement of $\mathbf{x}_1$ and $\mathbf{x}_2$ would guarantee a rectilinear motion between these points or that it would guarantee a momentum parallel to the line P_1P_2 are statements that cannot be verified experimentally, and that therefore better should not be made at all.

While, by a narrow hole (small $\Delta\mathbf{x}_2$), a state of sharply defined momentum (small $\Delta\mathbf{p}$) can neither be measured (see above) nor be prepared, one can prepare a state of sharply determined $\mathbf{p}$ by selecting, from the ensemble of energy K through narrow hole P_1 described by ψ_1 above, the subensemble that passes through a rather *wide* second hole (*large* $\Delta\mathbf{x}_2$) at sufficient distance from the first hole to be seen from it under a narrow solid angle (small $\Delta\mathbf{n}$, small $\Delta\mathbf{p}$). This *prepares* the desired state without conflict between ideas of preparation and of postparation.

APPENDIX C

ONE OF D'ESPAGNAT's "PROOFS" THAT THE RESULT OF A MEASUREMENT CANNOT BE A MIXED STATE

D'Espagnat has given a number of so-called "proofs" that the state after completion of a measurement could not possibly be reducible.[152] If these proofs would be correct, reduction of states after measurements would not be allowed, and this booklet should have remained unwritten.

The basic point in these proofs was d'Espagnat's insistence on a unitary time dependence of the density matrix. This point has been discussed by us on pages 19–20 and 29 where we have pointed out that this unitarity of evolution of the state in general does not hold for the measurement process, because the systems there considered usually are not closed, and usually depend upon randomly distributed external parameters, resulting in the occurrence of incoherence and in transitions from pure states to mixed states. (See Appendix D for an example.)

In this Appendix C, we mainly want to point out a misuse of postdiction made in two different versions of one of d'Espagnat's "proofs,"[134] if his reasoning is applied to more realistic measurements than he himself considered. In this particular "proof," by an integra-

[152] B. d'Espagnat, *Supplemento al Nuovo Cimento* **4**, 828 (1966). See also pages 288–294 of his 1971 book.[5]

tion of the Schrödinger equation backwards in time,[153] d'Espagnat tries to postdict, from an assumed mixed state at time t_2 at the end of a measurement, a state for the time t_1 at which the incident state was prepared, which is different from the state that at t_1 actually had been incident. The resulting contradiction according to d'Espagnat is to prove that after the measurement the state cannot have been reducible.

As we know from Chapter 1 that the latter conclusion is incorrect for successful measurements, there must be a flaw in d'Espagnat's reasoning. This flaw is easily identified. D'Espagnat's caricaturization of the measuring process is unadmissible for a successful measurement. He represents the measuring process by an interaction between the object and an apparatus, *assuming that the latter is entirely describable quantum-mechanically by a state vector*, with no interactions with a device that makes indelible recordings or with any other external system. It should be clear from this alone that the "measurement" discussed by d'Espagnat cannot be a successful one, and that the state at its end should not be expected to be reducible. (See Chapter 1. Especially, compare Sec. 1.20.)

If we want to leave it there, we may say that from there on d'Espagnat's reasoning is correct, and that the contradiction mentioned simply verifies that the measurement considered by d'Espagnat is incomplete (unsuccessful), as it does not lead to a mixed state.

As, however, d'Espagnat on his pages 287–288 suggests that he is going to show that the state of object and instrument together cannot be a mixed state after "measuring L," without saying that only unsuccessful measurements of L are contemplated, the reader will naturally assume that d'Espagnat's instrument does have some of the

[153] In this procedure, d'Espagnat never distinguishes postdictive states from predictive states. He simply reverses the time direction in which the Schrödinger equation for "the" state is integrated. However, d'Espagnat *does* recognize the difference in nature between predictive and postdictive knowledge elsewhere; in particular, at the top of his page 309, where he attacks a faulty reasoning presented by him on p. 308, on which he mistakenly calls this faulty reasoning "Bohr's philosophy." See footnote 137 on our page 88, and see the conclusion of this Appendix C.

"classical" properties that customarily are postulated for enabling the system of object with instrument to reach a final mixed state,[61, 68, 69, 70] and that may be described by external interactions or by including a device **M** that makes indelible records of the measurement results. From our Chapter 1, then, the reader will conclude that at the end of the measurement the ensemble of systems *is* in a reducible state in which object and instrument are separable, so that we may consider the ensemble of the objects alone and will find it in a mixed state, which is a probability distribution over pure object states each correlated to a corresponding memory state of the recording device. (See Secs. 1.21–1.24.) This allows us to consider reductions of state vectors (by introducing subensembles), and to use the conventional Copenhagen rules of quantum theory, and to use the methods of Chapter 2.

We then will ask ourselves: Where does d'Espagnat's reasoning break down, when we try to apply it to an ensemble of systems **O** interacting with effective measuring apparatus like the **AM** of Chapter 1, instead of with ineffective apparatus like **A** of Chapter 1 alone? Why does d'Espagnat's reasoning incorrectly describe what happens to the objects **O** when one traces them back from completion of successful measurements?

Looking closer at d'Espagnat's reasoning, we then will notice that d'Espagnat does not merely make a postdiction from t_2 back to t_1 (which we shall discuss below). He also assumes that (a) the reducible state at t_2 would be "*strictly* reducible," and (b) integration of the Schrödinger equation from t_2 backwards in time should at t_1 lead back to the same state as from which the state at t_2 was obtained by integration forwards in time.

Before we discuss the question of retrodiction from t_2 to t_1, let us first look into these two points (a) and (b), and see that d'Espagnat's reasoning, when applied to a *successful* measurement, would break down even if we would not notice that postdiction in this case is applied contrary to its rules of validity.

Point (a) assumes that, for validity of rule **F5** of page 16, quantum theory would assume "*strict* reducibility" of the state at the end of a measurement. By "*strict* reducibility" I mean that the density matrix would with mathematical rigor have lost all matrix elements between different measurement results. In practice, *strict* reducibility is merely a convenient idealization (or perhaps a formal limiting case for $t \to \infty$) of the effects of a measurement. These effects may be a result of *incoherence* after interaction with external systems, or they may be invoked as in Secs. 1.21–1.23 by the making of distinguishable *indelible records*.

In the *latter* case, assumption (b) ignores the irreversible nature of "making an indelible record." In the *former* case, assumption (b) ignores the nature of the processes by which the incoherence sets in that leads toward reducibility of the state at the end of the measurement. These are processes of randomization of phase factors by unpredictability of external factors influencing the evolution of the state by the Schrödinger equation. These processes are comparable to the ergodic processes of statistical mechanics, and to irreversible processes in thermodynamics. It is the main reason for the irreversibility of the measuring process, and for the time asymmetry of the probability interpretation of quantum mechanics. (See Chapter 2.) This randomization process leaves its mark whenever we use the Schrödinger equation on a system on which measurements are made, for calculating, from the state at one time, the state at a different time. In other words, there is randomization not only toward the future, but also toward the past, as in the H-theorem of kinetic theory, where it is shown that the entropy of a closed system at a different time would never be smaller than it is now, irrespective of whether that other time lies in the future or lies in the past.

The argument that integrating a differential equation first forward from t_1 to t_2, and then backward from t_2 to t_1, ought to lead back at t_1 to the initial status will certainly not be valid if at t_2, before the integration is reversed, we make a slight change in the quantity calculated. This is exactly what is done when with d'Espagnat we idealize the density matrix at t_2 by changing it from a practically reducible one into a strictly reducible one. But that is not all. During the integration from t_1 to t_2, randomization of phases took place because our ensemble is a mixture of subensembles differing from each other by microscopically different values of parameters that enter the integration of the Schrödinger equation. The vanishing at t_2 of the matrix elements between different measurement results was due to the destructive interference caused by this randomization. If upon integration back from t_2 to t_1 we would want the matrix elements that disappeared to reappear as they were originally, each of these subensembles should be integrated backward with in the Schrödinger equation the same values of these parameters as which for that particular subensemble were used on the integration from t_1 to t_2. Since, however, we have no ways of macroscopically distinguishing these subensembles from each other, and since the differences in the values of these parameters are too microscopic to observe, we have no way of performing the integration backward in this precisely time-inverted way. Therefore, we should not be shocked that, upon integrating the Schrödinger equation backward from the result of a measurement at t_2, to the time t_1 of the incident state, we do not find back the initial pure incident state, but we find a state that is at least as "garbled" as the state was already at t_2. There is nothing incomprehensible or paradoxical in this result, and this result does not cast any shadow over the assumption of a reducible state at t_2.

However this may be, our present attention is not for d'Espagnat's invalid "proofs" as such, but for a *misuse of retrodiction* contained in both versions of the one "proof" considered here, when we think it to be applied in connection with a successful measurement, that is, applied not to d'Espagnat's oversimplified systems **OA**, but to systems like **OAM** or thence to **O**.

Specifically, d'Espagnat in this "proof" discusses the special case of consecutive ideal Stern–Gerlach measurements[154] of $S(_x1)$ and $S_z(2)$ made upon an ensemble of silver atoms, where, as in Sec. 2.11, we indicated by numbers the time order in which these spin components are being measured.

[154] See footnote 87 on page 36 about the incoherence between the two beams occurring in a typical Stern–Gerlach measurement of spin direction.

ONE OF D'ESPAGNAT'S PROOFS

On pages 297–301 of his book,[5] d'Espagnat then tries to prove that in the subensemble E_{1+} [which is the one with $S_x(1) = +\hbar/2$] the spins of **O** at time t_2 could not be in the strictly reducible state which he imagines to be required for validity of the rule **F5**. He does so as follows.

If at time t_2 the ensemble of the **O** in E_{1+} interacting with the measuring instrument is in a strictly reducible mixed state, we may subdivide E_{1+} at that time into two subensembles, $E_{1+,2+}$ and $E_{1+,2-}$, corresponding to the two possible outcomes for $S_z(2)$. For each of these two subensembles of given value of $S_z(2)$, d'Espagnat now uses the Schrödinger equation backward in time[153] and concludes that at t_1 the objects **O** of both subensembles would be in a state of equal probabilities for $S_x(1) = +\hbar/2$ and for $S_x(1) = -\hbar/2$. The same would then also be true for their union, E_{1+}. Thus he finds for these probabilities in E_{1+} at time t_1 the values $P_+ = \frac{1}{2}$ and $P_- = \frac{1}{2}$, in *contradiction* to the fact that initially we had in E_{1+} the probabilities $P_+ = 1$ and $P_- = 0$.

D'Espagnat's blaming the reducibility of the state of $E_{1+}(t_2)$ for this contradiction shows another time that he did not consider realistic successful measurements of $S_z(2)$, for, in the above calculation, he ignored how, during integration back from t_2 to t_1, the external random influences on the apparatus for a real measurement would prevent the state calculated by the Schrödinger equation from losing the randomization which it had already acquired on its way forward from t_1 to t_2. Therefore his integration backward in time, starting from the correct mixed final state at t_2, for a realistic instrument *should be expected* to lead to a "postdicted" state at t_1 which shows more randomization (more "garbling") than the actual state initially incident at time t_1. The lack of equality between the postdicted state and the actual initial state then should not have been considered to be a contradiction, but rather a *confirmation* of the irreversibility of the random processes by which a realistic measuring instrument is affected.

If, however, preceding a successful measurement at t_2 one would want to find the correct values for the probabilities P_+ and P_- at t_1, one should use for that rules of quantum theory that are applicable for the purpose. For E_{1+}, the only way in which we can derive $P_+ = 1$ and $P_- = 0$ is by *predicting* it from the preceding measurement $S_x\,(1) = +\hbar/2$. The same probabilities then follow for its subensembles $E_{1+,\,2+}$ and $E_{1+,\,2-}$, which at t_1 are not yet distinguishable. Since we know from Sec. 2.6 [Eq. (43) on page 72] that postdiction is valid only in an ensemble that is initially garbled, and since E_{1+} and $E_{1+,\,2\pm}$ initially are *not* garbled, we *cannot use postdiction* for finding correct values for $P_\pm$.

Only for an *initially garbled* ensemble $E = E_{1+} \oplus E_{1-}$ in which $P_+ = P_-$ is given to start with, we could *postdict* for the time t_1 the probabilities P_+ and P_- from having at t_2 the ensemble E_{2+} $(= E_{1+,\,2+} \oplus E_{1-,\,2+})$ or E_{2-} $(= E_{1+,\,2-} \oplus E_{1-,\,2-})$. For these ensembles, postdiction would be allowed. For them, however, d'Espagnat's result $P_+ = P_- = \frac{1}{2}$ would not be very surprising.

Finding $P_\pm = \frac{1}{2}$ by (illegal) postdiction applied to $E_{1+,2\pm}$ merely *confirms* the unallowability of this postdiction.

Let us look again at what happens to the incident ensembles E_{1+} and E_{1-} separately, as they would travel through a *real* Stern-Gerlach apparatus.

In E_{1+} we have at t_1 one beam of particles polarized in the $+x$-direction. As they travel through the Stern–Gerlach apparatus for $S_z(2)$, the beam is split into two parts, $E_{1+,\,2+}$ and $E_{1+,\,2-}$. As the particles do not always enter the apparatus in geometrically exactly the same way, the two beams leave the apparatus with randomly distributed relative phases at time t_2. (See footnote 87 on page 36 for a numerical example.) At this time, the combined ensemble E_{1+} is practically in a mixed state.

In order to get back from this the initial state of E_{1+} at t_1 by integrating the Schrödinger equation backwards in time, we first would have to be able to split up the ensemble E_{1+} $(= E_{1+,\,2+} \oplus E_{1+,\,2-})$ at t_2 into subensembles each corresponding to definite values of the geometrical parameters that determine the phase shifts discussed.[155]

[155] The mathematical possibility of this different decomposition of $E_{1+}(t_2)$ is an illustration of the lack of uniqueness in interpretation of a mixed state as a probability distribution over pure states, discussed by us in Sec. 2.3.

In practice, however, it is *impossible* to split up $E_{1+}(t_2)$ in this way.[156] This split up is made impossible even *theoretically*, when the state of E_{1+} at t_2 is idealized as being *strictly* reducible.

Therefore, upon retracing the beams from t_2 to t_1, as we have only *macroscopic* information available on the beams in the apparatus, there is no process available for disentangling the mixing that took place on the way out (from t_1 to t_2). It is only natural that by attempts to retrace the beams we should find at t_1 an unpolarized state different from the actual initial polarized state of E_{1+}. There is nothing paradoxical in this result: It merely illustrates the irreversibility of the measuring process. This irreversibility is, of course, also responsible for the existence of the condition [Eq. (43) of page 72] for the validity of postdiction. Since E_{1+} does not satisfy this condition, we know from the start that postdiction is not possible on E_{1+}. Therefore, we should never even try to obtain the probability distribution for E_{1+} at t_1 from this distribution at t_2. If yet we postdict a state for $E_{1+}(t_1)$ from the state of $E_{1+}(t_2)$, we should not be puzzled by getting the wrong result. Getting the wrong result, we should recognize it as what we should have expected to find, and we certainly should not call it a proof that the state of E_{1+} at t_2 cannot have been reducible.

D'Espagnat may have felt uneasy about his result, because the formulas which he used in this first version would have been correct if he would have applied them to the ensembles $E_{2\pm}$ and E, for which $P_+ = P_- = \frac{1}{2}$ would *not* have yielded a contradiction. While he *said* that he was applying them to $E_{1+,\,2\pm}$ and E_{1+}, his formulas did not really show this.

Therefore, in sec. 15.5 of his book,[5] on pages 331–339 he added a second version, in which by small correction terms he took into account the effect of the initial angular momentum carried by the particles from the $S_x(1)$ apparatus, upon the $S_z(2)$ apparatus. Thus, his amended formulas did show a difference between the states of $E_{1+,\,2\pm}$, of $E_{1-,\,2\pm}$, and of their mixtures, $E_{2\pm}$ (E_{2+} and E_{2-}).

Again, he applied postdiction to the ensemble of the objects in

[156] Compare the top half of page 94.

$E_{1+,\ 2+}$ and in $E_{1+,\ 2-}$ interacting with the apparatus. (While his misrepresentation of this apparatus by an *undisturbed purely quantum-mechanical* system **A** made him think that this postdiction would be allowed, for any apparatus that can make a *successful* measurement this postdiction again is not allowed because of the lack of initial garbling.) For the resulting probabilities for $E_{1+} = E_{1+,\ 2+} \oplus E_{1+,\ 2-}$, he found now $P_+ \approx P_-$ instead of $P_+ = P_-$. In this, he inserted this time the initial probability $P_- = 0$, and thus erroneously found $P_+ \approx 0$ in disagreement with $P_+ = 1$. The source of the error is the same as before: In case of an undisturbed purely quantum-mechanical apparatus it is the assumption of a final mixed state, and for an apparatus that can make successful measurements it is the indiscriminate use of postdiction.

For the latter case, in the language of those who keep postdiction entirely out of the theory[157]: The results of a measurement cannot be used for "predicting the past." As this simple rule is what one would expect that Bohr's ciriticism would have been of d'Espagnat's attempts, it is incomprehensible why d'Espagnat on pages 340–343 claims that according to strict Copenhagen orthodoxy the imaginary problem raised by d'Espagnat should be solved, either by assuming that the second Stern-Gerlach apparatus does not measure S_z at all, or by assuming that the particles in the beam would already have[158] the values $S_z = \pm\hbar/2$ measured at t_2, *before* they reach the second Stern–Gerlach apparatus. It is even more amazing that he ascribes to Bohr[159] the idea that in the latter case the particles in the beam would take their S_z values at the instant when some observer would *plan* to measure these S_z by the Stern-Gerlach apparatus.[160] (Not only mind reading, but also clairvoyance.)

[157] See Sec. 2.12.

[158] The question should not be whether the particles "have" an S_z value, but whether they "have been prepared" to have an S_z value. This preparation, of course, takes place at t_2. (See our Secs. 1.10 and 2.11, and pages 101–102.)

[159] Reference 5, pages 173–175 and 404.

[160] Compare our footnote 137 on page 88.

APPENDIX D

EXAMPLE OF NONUNITARY EVOLUTION OF THE DENSITY MATRIX

Suppose a beam of spin $\frac{1}{2}$ particles in an $S_x = +\hbar/2$ state is incident upon a Stern–Gerlach apparatus measuring S_z. In a representation which diagonalizes S_z, the initial density matrix is

$$W_0 = \begin{pmatrix} \frac{1}{2} & \frac{1}{2} \\ \frac{1}{2} & \frac{1}{2} \end{pmatrix}.$$

The states $S_z = \pm\hbar/2$ will in the apparatus acquire phase factors which are sensitive to the exact distance traveled in the apparatus.[61] (For numerical values, see footnote 87 on page 36.) Let there be probabilities p_φ that states $S_z = \pm\hbar/2$ by passing through the apparatus acquire phase shifts by factors $\exp(\pm\frac{1}{2}i\varphi)$. Then, W_0 is changed into

$$\sum_\varphi p_\varphi \begin{pmatrix} \frac{1}{2} & \frac{1}{2}\exp(-i\varphi) \\ \frac{1}{2}\exp(i\varphi) & \frac{1}{2} \end{pmatrix}.$$

This is equivalent to probabilities p_φ for rotations of the incident spin by an angle φ in the xy-plane. If there are equal probabilities for all angles φ, the final density matrix becomes

$$\begin{pmatrix} \frac{1}{2} & 0 \\ 0 & \frac{1}{2} \end{pmatrix} = \frac{1}{2}\begin{pmatrix} 1 & 0 \\ 0 & 0 \end{pmatrix} + \frac{1}{2}\begin{pmatrix} 0 & 0 \\ 0 & 1 \end{pmatrix},$$

which may be interpreted as the reducible state in which there are 50%

probabilities for the states $S_z = \pm\hbar/2$. Therefore, no interference phenomena are destroyed, if we reduce this reducible ensemble by selecting the subensemble of all particles in one of the two beams.

This example differs from the case considered by d'Espagnat in his 1966 paper[49] in that *d'Espagnat admits only unitary transformations of* W, which of course leave the pure initial state W_0 pure at later times. That is: For him, φ does not have a probability distribution p_φ, but has the same value for the entire ensemble. This means that he assumes that both the hamiltonian and the time duration of the interactions shall be well determined and the same throughout the ensemble of measurements. It is impossible to satisfy these conditions rigorously in the laboratory.

APPENDIX E

CLOUD CHAMBER TRACKS WITHOUT REDUCTION OF STATES

When we consider the track of an alpha particle from a source at the origin, and for simplicity we think the vapor in the Wilson chamber to have been replaced by an array of electrons at fixed positions, the original spherical wave function ψ for the alpha particles, written in zeroth approximation as a spherical superposition of minimal-uncertainty gaussian wave packets traveling in all directions from the origin, will by the Schrödinger equation in first approximation develop terms describing slight scattering of these alpha particle wave packets with recoil of a first electron. In second approximation, each of these terms in ψ would develop additional terms describing a second scattering with recoil of a second electron. The terms which would describe the alpha particle together with two electrons recoiling from spots that do not approximately line up with the source of the alpha particle at the origin, however, would be found to appear with negligible coefficients in this expansion of ψ.

APPENDIX F

ON THE NEED FOR LARGE APPARATUS FOR VALIDITY OF THE CONVENTIONAL SCHEMATIZATION OF THE THEORY OF MEASUREMENTS

Wigner[7] has considered the case in which, on an object **O** by an apparatus **A**, a quantity A is measured while in the **OA** system a quantity B is conserved, with the corresponding operators $\mathcal{A}$ and $\mathcal{B}$ not commuting. If, for instance, in a Stern–Gerlach experiment we measure the spin component S_z of **O** first on the one and later on the other one of two incident beams of given values $+\hbar/2$ and $-\hbar/2$ of $S_x(\mathbf{O})$, the states of the upper as well as of the lower beam emerging from the S_z measurement may be described as superpositions of states in which $S_x(\mathbf{O})$ either is unchanged from the incident beam, or has flipped sign. The latter component of the final state, however, would have $S_x(\mathbf{O})$ *increased* by $\hbar$ if the incident beam was the one with $S_x(\mathbf{O}) = -\hbar/2$, and would have it *decreased* by $\hbar$ if its incident value was $+\hbar/2$. For conservation of angular momentum of the **OA** system, the apparatus should in its final "pointer position" states have components corresponding to a loss or to a gain of x-component of angular momentum, depending upon whether the incident particles had $S_x = -\hbar/2$ or $S_x = +\hbar/2$.

This dependence of the pointer position states of **A** upon the incident beam of **O** particles would be in contradiction to Eq. (3) of page 27, in which the apparatus states ξ_n depend merely upon the result A_n of the measurement, and not upon the coefficients c_n with which in Eq. (1) of page 27 the ϕ_n were linearly combined in the incident state of **O**. This contradiction, however, loses its importance when the apparatus is so big (with a "nearly infinite" x-component of angular momentum) that the loss or gain of a mere $\hbar$ makes practically no difference. (The accuracy with which the angular momentum of a macroscopic apparatus is known is not good enough for telling the difference.) When the apparatus is *not* that large, the difficulty could be removed, for instance, by some generalization of the theory, in which we would allow the states ξ_N of **A** to depend slightly upon the set of coefficients $\{c_n\}$ appearing in Eq. (1) [compare footnote 38 on page 16], as long as the differences in these apparatus states are small compared to the differences between the ξ_N for different N. Wigner[7] shows how consistency may also be achieved by adding to Eq. (3) of page 27 terms of the kind of

$$\sum c_n \, \gamma_{nml} \, \phi_m \, \xi_l',$$

where the ξ_l' are states of **A** that need not be recognizable as pointer positions at all. He shows that the coefficients γ_{nml} may be made small by increasing the size of the apparatus **A**.

APPENDIX G

DENSITY MATRICES AND MIXED STATES

Particularly Chapter 1 makes extensive use of the concepts of "density matrices" and "mixed states." As in some introductions to wave mechanics these are not sufficiently discussed, I will here give a brief explanatory outline.

The expected average of an observable A in a state ψ is assumedly

$$\langle A \rangle = \int \psi^*(x)\, \mathcal{A}\psi(x),$$

where $\mathcal{A}$ is the observable's operator in x-space. (By $\int, \int', \ldots$ we denote integration over all values of $x, x', \ldots$) Formally, we may write the operation of $\mathcal{A}$ in the form of an integral:

$$\mathcal{A}\, \psi(x) = \int' A(x, x')\, \psi(x').$$

For instance, $A(x, x')$ would be $= x\delta(x-x')$ if $\mathcal{A}$ is multiplication by x, and $A(x-x') = i\,\dfrac{\partial\delta(x-x')}{\partial x} = -i\,\dfrac{\partial\delta(x-x')}{\partial x'}$ if $\mathcal{A} = i\,\dfrac{\partial}{\partial x}$.

People often write $\langle x \,|\, A \,|\, x' \rangle$ for our $A(x, x')$. ⟨Dirac's notation.⟩

We define

$$\text{Trace}\quad \{\mathcal{A}\mathcal{B}\} = \int\!\!\int' A(x, x')\, B(x', x).$$

A state is called *pure*, if it can be described by a single state vector $\psi(x)$. The *density operator* $\mathcal{W}$ of a pure state is given by

$$W(x, x') = \psi(x)\, \psi^*(x').$$

Then, obviously,

$$\langle A \rangle = \int\!\!\int' \psi^*(x)\, A(x, x')\, \psi(x') = \int\!\!\int' A(x, x')\, W(x', x)$$
$$= \text{Trace}\,\{\mathcal{A}\mathcal{W}\} = \text{Trace}\,\{\mathcal{W}\mathcal{A}\}.$$

If $\{\phi_i(x)\}$ is a complete orthonormal set of functions, we may write

$$\psi(x) = \sum c_i \, \phi_i(x), \qquad A_{ij} = \int \phi_i^*(x) \, \mathcal{A} \, \phi_j(x),$$
$$\langle A \rangle = \sum \sum c_i^* \, A_{ij} \, c_j = \text{Trace}\,\{\mathcal{A}\mathcal{W}\} = \sum \sum A_{ij} W_{ji},$$
$$W_{ij} = c_i \, c_j^*, \qquad W(x, x') = \sum \sum \phi_i(x) \, W_{ij} \, \phi_j^*(x').$$

$W(x, x')$ and W_{ij} are called the *density matrix* in the x representation and in the i representation, respectively.

States are not always pure. Consider, for instance, an ensemble of ordinary cats. Of them, let 36% be dead and let 64% be living. Though we don't know *which* of the cats are still living, we *do* know that each cat either *is* alive or *is* dead. That is, we *may* talk about cats in the language of classical physics. If, in quantum-mechanical language, ψ_L is the state vector describing cats that are alive, and ψ_D describes dead cats, the ensemble of cats here considered has a probability $p_L = 0.64$ for ψ_L and a probability $p_D = 0.36$ for ψ_D. Therefore, if A is some catty observable, its average over the ensemble is

$$\langle A \rangle = p_L \int \psi_L^* \, \mathcal{A} \, \psi_L + p_D \int \psi_D^* \, \mathcal{A} \, \psi_D = \text{Trace}\,\{\mathcal{A}\mathcal{W}\},$$

where this time we have defined the density matrix $\mathcal{W}$ by

$$W(x, x') = p_L \, \psi_L(x) \, \psi_L^*(x') + p_D \, \psi_D(x) \, \psi_D^*(x')$$

as a linear combination of the density operators for the alive state and for the dead state. This kind of a classical probability distribution over states is called a *mixed state*.

For the product of two operators, $\mathcal{C} = \mathcal{A}\mathcal{B}$, obviously

$$C(x, x') = \int'' A(x, x'') \, B(x'', x').$$

Normalization of ψ gives *for each pure state*

$$\int'' W(x, x'') \, W(x'', x') = \psi(x) \left[\int'' \psi^*(x'') \, \psi(x'')\right] \psi^*(x')$$
$$= \psi(x) \, \psi^*(x') = W(x, x'),$$

so that $\mathcal{W}^2 = \mathcal{W}$ *for pure states.*

For mixed states, however, $\mathcal{W}^2 \neq \mathcal{W}$. For instance, for the above cats, orthonormalization of ψ_L and ψ_D gives

$$\begin{aligned}\int'' W(x, x'')\, W(x'', x') &= p_L^2\, \psi_L(x) \left[\int'' \psi_L^*(x'')\, \psi_L(x'')\right] \psi_L^*(x') \\ &+ p_L p_D \left\{\psi_L(x) \left[\int'' \psi_L^*(x'')\, \psi_D(x'')\right] \psi_D^*(x') + \text{conjugate}\right\} \\ &+ p_D^2\, \psi_D(x) \left[\int'' \psi_D^*(x'')\, \psi_D(x'')\right] \psi_D^*(x') \\ &= p_L^2\, \psi_L(x)\, \psi_L^*(x') + p_D^2\, \psi_D(x)\, \psi_D^*(x').\end{aligned}$$

Therefore, $\mathcal{W} = p_L \mathcal{W}_L + p_D \mathcal{W}_D$ with $\mathcal{W}^2 = p_L^2 \mathcal{W}_L + p_D^2 \mathcal{W}_D$ gives $\mathcal{W}^2 \neq \mathcal{W}$ by $p_L^2 < p_L$ and $p_D^2 < p_D$ whenever the state is not pure.

The pure states $\psi_L, \psi_D, \ldots$ and their probabilities $p_L, p_D, \ldots$ unambiguously determine the density matrix $W(x, x') = \sum p_n \psi_n(x) \psi_n^*(x')$ of a mixed state. Conversely, a given density matrix may in infinitely many ways be written as a probability distribution over pure states. (For some examples, see pages 58–61.) As $\mathcal{W}$ must be hermitian, one possibility which always exists is taking for the ψ_n the eigenfunctions of $\mathcal{W}$, and for the p_n the corresponding eigenvalues W_n. Then $W_n \psi_n = \mathcal{W} \psi_n$ gives $\sum W_n \psi_n(x)\, \psi_n^*(x') = \sum \int'' W(x, x'')\, \psi_n(x'')\, \psi_n^*(x')$ $= \int'' W(x, x'') \delta(x'' - x') = W(x, x')$ by the completeness of the orthonormal set $\{\psi_n(x)\}$.

If all states are transformed unitarily according to $\psi' = \mathcal{U}\psi$ (with $\mathcal{U}^*\mathcal{U} = 1$, if $\mathcal{U}^*$ is the hermitian conjugate of $\mathcal{U}$), we would transform the ψ^* into $\psi^*\mathcal{U}^*$, and density matrices by $\mathcal{W}' = \mathcal{U}\mathcal{W}\mathcal{U}^*$. Since then $\mathcal{W}'^2 - \mathcal{W}' = \mathcal{U}(\mathcal{W}^2 - \mathcal{W})\mathcal{U}^*$, it follows that after unitary transformation $\mathcal{W}'^2 = \mathcal{W}'$ if $\mathcal{W}^2 = \mathcal{W}$, and $\mathcal{W}'^2 \neq \mathcal{W}'$ if $\mathcal{W}^2 \neq \mathcal{W}$, so that pure states go into pure states, and mixed states go into mixed states. If the time integral of the hamiltonian is the same for all ψ occurring with various probabilities in a mixed state, the unitary operator $\mathcal{U}(t)$ $= \exp\left[\int \mathcal{H} dt / i\hbar\right]$ by which all ψ then would evolve according to $\psi(t) = \mathcal{U}(t)\, \psi(0)$ will transform pure states into pure states, and mixed states into mixed states. If, however, we mix states for which $\int \mathcal{H}\, dt$ differs from one ψ to another, there is no longer a unitary transformation for the density matrix as a whole, and transitions between pure states and mixed states will take place.

AUTHOR INDEX

SUBJECT INDEX

OTHER TITLES IN THE SERIES IN NATURAL PHILOSOPHY

Vol. 1. Davydov—Quantum Mechanics
Vol. 2. Fokker—Time and Space, Weight and Inertia
Vol. 3. Kaplan—Interstellar Gas Dynamics
Vol. 4. Abrikosov, Gor'kov and Dzyaloshinskii—Quantum Field Theoretical Methods in Statistical Physics
Vol. 5. Okun'—Weak Interaction of Elementary Particles
Vol. 6. Shklovskii—Physics of the Solar Corona
Vol. 7. Akhiezer *et al.*—Collective Oscillations in a Plasma
Vol. 8. Kirzhnits—Field Theoretical Methods in Many-Body Systems
Vol. 9. Klimontovich—The Statistical Theory of Non-equilibrium Processes in a Plasma
Vol. 10. Kurth—Introduction to Stellar Statistics
Vol. 11. Chalmers—Atmospheric Electricity (2nd Edition)
Vol. 12. Renner—Current Algebras and their Applications
Vol. 13. Fain and Khanin—Quantum Electronics, Volume 1—Basic Theory
Vol. 14. Fain and Khanin—Quantum Electronics, Volume 2—Maser Amplifiers and Oscillators
Vol. 15. March—Liquid Metals
Vol. 16. Hori—Spectral Properties of Disordered Chains and Lattices
Vol. 17. Saint James, Thomas and Sarma—Type II Superconductivity
Vol. 18. Margenau and Kestner—Theory of Intermolecular Forces (2nd Edition)
Vol. 19. Jancel—Foundations of Classical and Quantum Statistical Mechanics
Vol. 20. Takahashi—An Introduction to Field Quantization
Vol. 21. Yvon—Correlations and Entropy in Classical Statistical Mechanics
Vol. 22. Penrose—Foundations of Statistical Mechanics
Vol. 23. Visconti—Quantum Field Theory, Volume 1
Vol. 24. Furth—Fundamental Principles of Modern Theoretical Physics
Vol. 25. Zheleznyakov—Radioemission of the Sun and Planets
Vol. 26. Grindlay—An Introduction to the Phenomenological Theory of Ferroelectricity
Vol. 27. Unger—Introduction to Quantum Electronics
Vol. 28. Koga—Introduction to Kinetic Theory: Stochastic Processes in Gaseous Systems
Vol. 29. Galasiewicz—Superconductivity and Quantum Fluids
Vol. 30. Constantinescu and Magyari—Problems in Quantum Mechanics
Vol. 31. Kotkin and Serbo—Collection of Problems in Classical Mechanics
Vol. 32. Panchev—Random Functions and Turbulence
Vol. 33. Talpe—Theory of Experiments in Paramagnetic Resonance
Vol. 34. ter Haar—Elements of Hamiltonian Mechanics (2nd Edition)
Vol. 35. Clarke and Grainger—Polarized Light and Optical Measurement
Vol. 36. Haug—Theoretical Solid State Physics, Volume 1
Vol. 37. Jordan and Beer—The Expanding Earth
Vol. 38. Todorov—Analytical Properties of Feynman Diagrams in Quantum Field Theory

Vol. 39. SITENKO—Lectures in Scattering Theory
Vol. 40. SOBEL'MAN—An Introduction to the Theory of Atomic Spectra
Vol. 41. ARMSTRONG and NICHOLLS—Emission, Absorption and Transfer of Radiation in Heated Atmospheres
Vol. 42. BRUSH—Kinetic Theory, Volume 3
Vol. 43. BOGOLYUBOV—A Method for Studying Model Hamiltonians
Vol. 44. TSYTOVICH—An Introduction to the Theory of Plasma Turbulence
Vol. 45. PATHRIA—Statistical Mechanics
Vol. 46. HAUG—Theoretical Solid State Physics, Volume 2
Vol. 47. NIETO—The Titius-Bode Law of Planetary Distances: Its History and Theory
Vol. 48. WAGNER—Introduction to the Theory of Magnetism
Vol. 49. IRVINE—Nuclear Structure Theory
Vol. 50. STROHMEIER—Variable Stars
Vol. 51. BATTEN—Binary and Multiple Systems of Stars
Vol. 52. ROUSSEAU and MATHIEU—Problems in Optics
Vol. 53. BOWLER—Nuclear Physics
Vol. 54. POMRANING—The Equations of Radiation Hydrodynamics
Vol. 55. BELINFANTE—A Survey of Hidden Variables Theories
Vol. 56. SCHEIBE—The Logical Analysis of Quantum Mechanics
Vol. 57. ROBINSON—Macroscopic Electromagnetism
Vol. 58. GOMBÁS and KISDI—Wave Mechanics and Its Applications
Vol. 59. KAPLAN and TSYTOVICH—Plasma Astrophysics
Vol. 60. KOVÁCS and ZSOLDOS—Dislocations and Plastic Deformation
Vol. 61. AUVRAY and FOURRIER—Problems in Electronics
Vol. 62. MATHIEU—Optics
Vol. 63. ATWATER—Introduction to General Relativity
Vol. 64. MULLER—Quantum Mechanics: A Physical World Picture
Vol. 65. BILENKY—Introduction to Feynman Diagrams
Vol. 66. VODAR and ROMAND—Some Aspects of Vacuum Ultraviolet Radiation Physics
Vol. 67. WILLETT—Gas Lasers: Population Inversion Mechanisms
Vol. 68. AKHIEZER—Plasma Electrodynamics, Volume 1—Linear Theory
Vol. 69. GLASBY—The Nebular Variables
Vol. 70. BIAŁYNICKI-BIRULA—Quantum Electrodynamics
Vol. 71. KARPMAN—Non-linear Waves in Dispersive Media
Vol. 72. CRACKNELL—Magnetism in Crystalline Materials
Vol. 73. PATHRIA—The Theory of Relativity
Vol. 74. SITENKO and TARTAKOVSKII—Lectures on the Theory of the Nucleus
Vol. 75. BELINFANTE—Measurements and Time Reversal in Objective Quantum Theory
Vol. 76. SOBOLEV—Light Scattering in Planetary Atmospheres
Vol. 77. NOVAKOVIĆ—The Pseudo-spin Method in Magnetism and Ferroelectricity
Vol. 78. NOVOZHILOV—Introduction to Elementary Particle Theory
Vol. 79. BUSCH and SCHADE—Lectures on Solid State Physics